A large part of the effort of the food industry is spent in attempting to understand the limitations of the type of food that animals can eat. An understanding of the factors that influence feeding behaviour can then be used to produce foodstuffs that are more attractive to the animal in question, whether it be man, cattle, dogs or cats. This book examines both the texture of food and the adaptations of various animals (including fish, mammals, primates and man) to the type of food they commonly eat. Zoologists, material scientists and food scientists have come together to present for the first time an integrated overview of feeding by vertebrates. The mechanical properties of various foods are considered in conjunction with the mechanics of eating them and more subjective behavioural parameters such as acceptability and palatability. The book consequently will be of interest to food scientists, zoologists and animal behaviourists.

SOCIETY FOR EXPERIMENTAL BIOLOGY
SEMINAR SERIES: *44*

FEEDING AND THE TEXTURE OF FOOD

SOCIETY FOR EXPERIMENTAL BIOLOGY SEMINAR SERIES

A series of multi-author volumes developed from seminars held by the Society for Experimental Biology. Each volume serves not only as an introductory review of a specific topic, but also introduces the reader to experimental evidence to support the theories and principles discussed, and points the way to new research.

Contents

Contributors

CORLETT, R. T.
Department of Botany, Pokfulam Road, The University, Hong Kong
EVES, A.
Leatherhead FRA, Randalls Road, Leatherhead, Surrey KT22 7RY, UK
HEATH, M. R.
Department of Prosthetic Dentistry, The London Hospital Medical College Dental School, Turner Street, London E1 2AD, UK
JERONIMIDIS, G.
Biomechanics Group, Department of Engineering, The University, Whiteknights, PO Box 225, Reading RG6 2AY, UK
KILCAST, D.
Leatherhead FRA, Randalls Road, Leatherhead, Surrey KT22 7RY, UK
LILLFORD, P. J.
Unilever Research, Colworth House, Sharnbrook, Bedford MK44 1LQ
LUCAS, P.
Anatomy Department, National University of Singapore, 10 Kent Ridge Crescent, Republic of Singapore, 0511 Singapore
OTTEN, E.
Department of Anatomy, Rijksuniversiteit Groningen, 9712 KZ Groningen, Bloemsingel, The Netherlands
PURSLOW, P. P.
AFRC Institute of Food Research, Bristol Laboratory, Langford, Bristol BS18 7DY, UK
Present address, Department of Veterinary Medicine, School of Veterinary Science, Churchill Building, Langford, Bristol, BS18 7DY, UK
RODGER, G. W.
Marlow Foods, PO Box 1, Billingham, Cleveland TS23 1LB, UK
SIBBING, F. A.
Department of Experimental Morphology, PO Box 338, 6700 AH Wageningen, The Netherlands

SMITH, A.
AFRC Institute of Food Research, Norwich Laboratory, Colney Lane, Norwich NR4 7UA, UK
VINCENT, J. F. V.
Biomechanics Group, Department of Pure and Applied Zoology, The University, Whiteknights, PO Box 228, Reading RG6 2AJ, UK

J. F. V. VINCENT AND P. J. LILLFORD

Introduction

To the biologist, especially the zoologist, the mechanical properties of food are treated rather vaguely: from an early age one is taught that carnivorous mammals have special teeth for coping with meat, but the properties of the meat which is being coped with are left variously as vague, assumed or obvious. The real answer is not only that the mechanical properties of meat are very subtle and difficult to measure, but that they are variably dependent upon a range of factors, few of which are properly understood (Purslow, this volume). Problems can be taken to a finer degree of resolution: within the range of animals which eat other animals it is possible to differentiate between the dentition of those which are specialised to eat hard-bodied insects (e.g. beetles) and those which can eat soft-bodied insects (e.g. caterpillars). Recent work (J.F.V. Vincent and S. Strait, unpublished results) suggests that the differences in dentition in these two types of animal (which may be closely related bats, or primitive primates) are based on the fracture mechanics of the food items – beetles are brittle and can be crushed; caterpillars are more ductile and have to be cut.

Food scientists are aware of some of these subtle differences in the mechanical properties of food and their implications. Much current research in industry and academia is directed at understanding the mechanics of food and the ways in which the mechanical properties of a wider variety of starting materials can be brought into the relatively narrow area of mechanical competence of the mouth. This can be illustrated by considering man's early efforts in food science.

Chewing food is a process of fracture, breaking the food into particles which are small enough to be swallowed. Typically this is of the order of a 1 mm cube. Fracture, a process requiring energy (Jeronimidis, this volume) is, in its turn, limited during chewing by the amount of strain energy available from the jaw muscles (Otten and Kilcast, both this volume). It follows that a compliant food, if it cannot be strained to failure, will register as 'tough' – i.e. unbreakable, even though the jaw muscles can supply sufficient force to break the food item. Foods which will fracture at low strains are called variously 'tender', 'crisp' or 'brittle' (depending on the forces required for fracture), independently of such descriptors as 'hard' or 'springy' which relate to stiffness or to the ease of starting a crack (Lillford, this volume).

Man probably evolved as an eater of nuts and fruit. *Australopithecus boisei* was given the nickname 'Nutcracker Man' on account of his dentition, though he was by no means unique in this respect. Examination of his jaw mechanics suggests that *A. boisei* was probably no more capable of cracking a tough or hard nut than we are (Leakey, 1981, p. 71) and was similarly limited in the deformations or strains which he could impose upon food in his mouth when eating. One could then also make the proposition that foods which modern man (and presumably his forebears) will eat 'raw' could have constituted a primitive diet. Such foods are fruits, nuts, seeds, vegetative parts of some plants (e.g. tubers; non-fibrous fleshy leaves, petioles and stems; shoots), all of which are, or can be, tender, crisp or brittle (Lucas and Corlett, this volume). Fish also comes under this heading (Rodger, this volume). That such textures are (within limits) the most 'acceptable' is surely no coincidence! Some foods (most meats, fibrous plants, starchy seeds) fall outside this textural range, being categorised as 'tough' or 'hard', indicating that the human mouth cannot generate the requisite strain (or stress, which can still be limiting) to fracture them. These are generally the foods which are cooked. The process of cooking these foods brings their mechanical properties within the range with which the mouth can cope.

It is significant that the earliest remains of 'man' (as opposed to an advanced ape) are associated with the earliest evidence of manipulation of fire. Thus the introduction of 'cooking' (in its broadest sense), made possible with the taming of fire, brought a wider range of foods to man and thereby increased his ecological and evolutionary fitness and therefore dominance. Obviously cooking also has chemical influence, such as gelatinising starches and making them more digestible (Smith, this volume) and denaturing or destroying toxic chemicals and microorganisms. Whilst the chemical effects can sometimes be correlated with a visual one (e.g. browning of meat), the mere fact that assessment of the completion or thoroughness of cooking is achieved by mechanical tests (e.g. prodding a stick into the object: is it 'done'?; does it fall apart more readily? has it 'set'?) illustrates the importance of cooking as a modifier of mechanical properties. The effects of cooking on the mechanical properties of food appear to have been designed to make the mechanics of the food more appropriate to the abilities of the mouth to disrupt it. 'Tenderising' is due partly to the dissolution, at high temperature, of chemicals and tissues holding together cells and fibres (e.g. collagen in meat; central lamella between cells in plants), whilst baking produces brittle foods partly by precipitation and denaturation of the ingredients and partly by the removal of water.

The chapters in this volume, originally presented at the SEB meeting in Edinburgh, April 1989, expand on the above ideas, proceeding from mechanical properties in general, the mechanics of food and feeding in a variety of foods and animals, through to the current industrial and research view of our own food. The editors hope that you will concur with their view – that food and feeding is an area where there is much mutual interest between biologists and industry.

Reference

Leakey, R.E. (1981). *The Making of Mankind.* Rainbird Publishing Group, London.

G. JERONIMIDIS

Mechanical and fracture properties of cellular and fibrous materials

Most foods, natural or manufactured, have structures which are either cellular or fibrous or both. Meat, fish and poultry owe their distinctive texture to the fibrous nature of the muscles and to the way in which they fracture with fibre separation. The crispness of fresh fruit and vegetables is a consequence of having moderately large cells filled with liquid, pressurised by osmotic turgor and adhering to each other. On being bitten or chewed the cells can either rupture, releasing their contents as in a crisp apple, or separate from each other as in a mealy apple. When significant amounts of fibres or fibre bundles are also present, as in asparagus or spinach for example, the texture of the cooked vegetable is affected considerably and the material is perceived as tough because of the difficulty in breaking the fibres. Extrusion-cooked products are often designed to be heterogeneous so as to introduce desirable textural attributes; this can be achieved, for example, by producing cellular structures where cell size and cell wall thickness can be varied.

The perception of food texture depends on specific properties of the food materials themselves (resistance to deformation and fracture, appearance, taste, etc.) and on the interactions between the mouth and the foods through teeth, muscles, taste buds, etc. (Williams & Atkin, 1983). Biting and mastication in particular play a central role in determining the acceptability of foods and this chapter concentrates on the mechanical properties of food materials relevant to these aspects. The elastic, strength and fracture properties of heterogeneous substances will be reviewed and discussed in relation to those textural attributes which depend on them.

Mouth–food interactions

When food is placed in the mouth and biting or chewing takes place, the teeth apply a *deformation* to the food material through the action of the muscles. The distinction between the application of a deformation and that of a load is important because in many instances the stability of cracks depends on which variable is controlled (Atkins & Mai, 1985) and this, in turn, has an effect on the choice of test methods which need to be used for proper characterisation of fracture properties.

As a result of the applied deformation, the food material responds elastically or viscoelastically at first, depending on the rate of application of the deformation, until some critical strain value is reached when either flow or fracture or a combination of both will take place. The *stresses* induced in the food materials will depend on the geometry and size of piece put in the mouth, on its elastic or viscoelastic properties and, in the case of anisotropic materials such as meat, also on the orientation of the piece with respect to the teeth.

The stresses associated with biting and chewing are seldom simple states of stress such as uniaxial tension or compression; heterogeneous food materials are particularly sensitive to complex stress states. The existence of multiple failure mechanisms at different levels of structure leads to several failure modes, each associated with a particular stress or combination of stresses, increasing the complexity of characterisation and analysis. For an isotropic substance which fails by plastic flow, for example, the yield strength is the only strength parameter needed because tensile, compressive or shear strength can all be associated with a single yield strength value. Even the simplest fibrous composite, on the other hand, requires five strength parameters for full characterisation (Jones, 1975).

When food materials are processed in the mouth the types of failure which are associated with comminution can be subdivided into four broad categories:

> brittle (boiled sweets, chocolate, hard biscuits, nuts, etc.),
>
> ductile (soft cheeses, soft toffees, etc.),
>
> ductile-brittle (chocolate, hard cheeses, etc.),
>
> fibrous (meat, fish, etc.).

Cellular materials, liquid-filled or not, can be assigned to either of the first three categories depending on cell size, properties of cell wall materials and adhesive strength between cells. In many manufactured foods such as air-filled chocolate products and starch-based extrusion-cooked materials, pseudo-ductile behaviour at the macroscopic or mouth level can be incorporated in the design, although the failure is essentially brittle at the microscopic level. In apples, for example, the specific design of the cellular structure can produce brittle or tough fractures during the first bite, depending on the type of apple and on orientation in the mouth (Khan, 1989). This is an illustration of how important it is to relate the level of heterogeneity of a given food material to the absolute size and shape of the volume introduced into the mouth for comminution. A proper understanding of mouth–food interactions in relation to texture must consider these aspects especially for predictive and design purposes.

All materials are heterogeneous at some level of structure or another and the mechanical properties that are measured are the result of some averaging process over a given volume. Foods vary considerably in their heterogeneity, both in size and in kind, and this does affect those textural attributes which depend on measured mechanical properties. The definition of a *representative volume* and of the rules to apply to it for averaging purposes is essential. The various failure modes defined earlier are not absolute but depend on these scale effects. A brittle material may deform in a ductile manner if sufficiently small and, conversely, a ductile material can fail in a brittle manner if sufficiently large. The difficulty lies in defining what 'sufficiently' means but there are experimental and theoretical techniques available for the purpose (Hashin, 1962; Mai & Atkins, 1980).

When dealing with food materials the effects of water need also to be taken into account. Water can act as a plasticiser (Lillford, 1988), altering properties of the base materials, or play a more direct structural role as in liquid-filled cellular foods such as fresh foods and vegetables (Jeronimidis, 1988; Gibson & Ashby, 1988). In both instances the presence of water can alter the type of failures produced by mouth action, changing a food material from brittle to ductile or pseudo-ductile, with associated differences in perceived texture. In this respect it is particularly important to consider the water pick-up of the food *after* introduction into the mouth, the rate of pick-up and its short-term effects on elastic and fracture properties.

Heterogeneous materials: structure and properties overview

In dealing with heterogeneity and mechanical properties of food materials it is convenient to use definitions and descriptions already established for other classes of materials (Cottrell, 1964). This is because most of the theoretical and experimental techniques have been developed in that context and have led to unambiguous definitions of properties which are useful in avoiding confusion. Terms such as stiffness, strength, fracture, ductility, etc. have very precise meaning in engineering and materials science, whereas, all too often, the corresponding terminology associated with food textural studies is less precise: crispness, crunchiness, firmness, etc. Indeed, part of the development of food mechanics as a discipline will require the establishment of appropriate methodologies to relate the engineering terms to those of the food textural studies.

From the point of view of their physical state all materials, including foods, can be subdivided into solids, liquids and gases. The distinction between the last two is straightforward but the same is not true for the first two, especially in relation to mechanical properties, viscoelastic behaviour and water effects. Depending on the rate of deformation, the temperature and the amount of water pick-up, the same material can be looked upon as a solid or as a more or less viscous liquid. The same piece of chocolate will fracture into several smaller bits when cold and deformed quickly, exhibiting all the characteristics of a brittle solid, but will flow irreversibly as a liquid if the temperature is sufficiently high or the rate of deformation sufficiently low. As mentioned earlier, however, even a cold piece of chocolate loaded rapidly can yield and flow if it is very small.

Within the materials which under the given circumstances behave like solids, a further distinction needs to be made between isotropy and anisotropy. This affects the elastic as well as the strength and fracture properties and will be discussed later in greater detail. Anisotropic food materials owe their anisotropy to their multiphase composition. Characteristic aspect ratios and linear dimensions of particles, voids and fibres with respect to representative volume determine not only whether or not the material should be considered homogeneous but also whether or not it should be considered anisotropic. If, for example, a manufactured food contains fibres which are too small to be perceived by the teeth and locally aligned, the material is strictly speaking microscopically heterogeneous and anisotropic; but if the volume under consideration is much greater than the representative volume it will behave macroscopically as a homogeneous isotropic material. Conversely if such a material needs to be redesigned to introduce specific textural attributes associated with heterogeneity and anisotropy, the phase dimensions will need to be changed or higher levels of structure will need to be introduced during processing.

Isotropic and anisotropic food materials can further be subdivided according to the type and level of deformation required to initiate flow or fracture. Some foods, which are best described as rubbery, will deform reversibly up to comparatively large strains (10% or more). Given the limited level of deformation that mouth and teeth can apply it may not be possible to reach the strain levels at which fracture will occur and they may well be perceived as tough, although they may not be tough at all in an engineering sense. In this case comminution in the mouth requires a specific cutting action from the teeth. The whole process is like trying to bite into or chew a rubber elastic band; rubber is a fairly brittle material. This distinction between low strain and high strain fracture is also relevant to the behaviour of gas-filled cellular structures as demonstrated, for example, by the differences in compressibility and compressive strength between fresh and stale bread.

The techniques which need to be used to study the mechanical properties of food material are no different conceptually from those already established with other types of materials, but difficulties do arise both experimentally and theoretically. Many foods cannot be obtained in large enough sizes or regular enough geometries to satisfy the validity requirements that certain types of mechanical test demand. As an example of this, consider a simple three-point bending experiment to determine Young's modulus or bending strength (Fig. 1). If the material is isotropic and homogeneous, even a small sample will give accurate results so long as the span is about ten times the thickness. If on the other hand the material is moderately anisotropic and heterogeneous, the span to thickness ratio will have to be 100 or more before sensible results are obtained (Fellers & Carlsson, 1979). Similar considerations apply to fracture testing as mentioned earlier.

Theoretical difficulties arise from the complexity of anisotropic and heterogeneous food materials, especially in relation to the predictions of *effective properties* as the average of complex interactions at specific levels of structure within appropriate representative volumes. Micro- and macromechanics techniques used successfully with artificial fibrous composites and cellular solids are available but the limits of applicability need

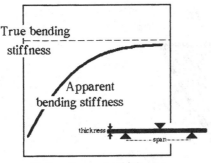

Ratio of span to thickness

Fig. 1. The effect of changing the span-to-thickness ratio on the measurement of stiffness in bending in an anisotropic material.

to be established and assessed for each particular case (Ashby, 1983; Hull, 1981). A particular problem exists with high strain anisotropic materials, about which little is known even for much simpler materials than natural or artificial foods (Jeronimidis & Vincent, 1984).

Fibrous materials

Elastic properties

The elastic properties of fibrous composites are derived from micromechanics principles which take into account the interactions between the various phases. The simplest conceptual model consists of parallel, infinitely long fibres embedded in a 'matrix' of lower stiffness and lower strength than the fibres (Fig. 2). Knowing the elastic properties of fibres and matrix, the properties of the composite are obtained from first principles (Jones, 1975).

If the material is loaded in tension or compression parallel to the fibre direction and if the elastic strains in both phases are assumed to be equal, the composite Young's modulus E, in the fibre direction is given by the 'rule of mixtures':

$$E_\ell = E_f V_f + (1-V_f) E_m \tag{1}$$

where the subscripts f and m relate to fibres and matrix, respectively, and ℓ stands for longitudinal, i.e. parallel to fibres. V_f is the volume fraction of fibres in the system.

When the tensile or compressive load is applied normal to the fibre direction (t for transverse) the fibres and the matrix are assumed to carry the same stress and the transverse Young's modulus is given by:

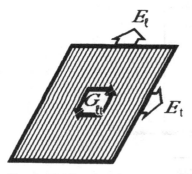

Fig. 2. Unidirectional thin composite lamina – principal directions of anisotropy. E, Young's modulus; G, shear modulus; subscripts ℓ, t, longitudinal, transverse.

$$E_t = \frac{E_f E_m}{E_f V_m + E_m V_f} \qquad (2)$$

A similar expression is obtained for the 'in plane' shear modulus, $G_{\ell t}$, of the material:

$$G_{\ell t} = \frac{G_f G_m}{G_f V_m + G_m V_f} \qquad (3)$$

By the same type of argument the principal Poisson's ratio can also be derived from the corresponding values for fibres and matrix:

$$v_{\ell t} = v_f V_f + v_m V_m \qquad (4)$$

This Poisson's ratio corresponds to the *induced strain* ε_t in the t direction for an *applied strain* ε_ℓ in the longitudinal direction and is defined as:

$$v_{\ell t} = -\frac{\varepsilon_t}{\varepsilon_\ell} \qquad (5)$$

There is also a 'minor' Poisson's ratio, $v_{t\ell}$, which measures the induced longitudinal strain for a given applied transverse strain.

Four out of these five elastic constants are independent and need to be measured accordingly. The minor Poisson's ratio is related to E_ℓ, E_t and $v_{\ell t}$ by:

$$\frac{v_{\ell t}}{E_\ell} = \frac{v_{t\ell}}{E_t} \qquad (6)$$

Even the simplest fibrous composite requires twice as many independent elastic properties as an isotropic material where E and v or E and G or G and v are sufficient for full characterisation.

This analysis applies to thin composite layers where the thickness can be considered small compared to the lateral dimensions of the material. When this is not the case a full set of either five or nine independent elastic properties is needed, depending on whether the material possesses a plane of isotropy or not. In theory these properties can be calculated from expressions as those given here but their predictive power is often very limited. The only expressions which have been found to work reasonably well are those for the Young's modulus in the fibre direction and for the principal Poisson's ratio; in the other cases one needs to resort to experiment (Tsai, 1987).

When the fibres have a finite length and are not parallel to each other the expressions given earlier need to be modified to take these factors into account. Equation (1), for example, can be re-written as:

$$E_\ell = \alpha\beta E_f V_f + (1 - V_f) E_m \qquad (7)$$

where α is a parameter dependent on fibre length and β a parameter dependent on fibre orientation. Expressions for these parameters have been calculated (Cox, 1952):

$$\alpha = 1-[\tanh (\tfrac{1}{2} k\ell) / \tfrac{1}{2} k\ell] \tag{8}$$

where ℓ is the fibre length and k a parameter depending on volume fraction of fibres, shear modulus of matrix and Young's modulus of fibres. As ℓ increases α approaches the value of 1.

$$\beta = \int_{0}^{\pi} \cos^4 \theta \, f(\theta) \, d\theta \tag{9}$$

where $f(\theta)$ is a function describing the statistical orientation of fibres. When the fibre orientation is random in the plane, $\beta = 1/3$. When it is random in three-dimensional space, $\beta = 1/6$ and when the fibres are all parallel, $\beta = 1$. For orientations which are intermediate between parallel and random, the distribution function of fibre orientations must be obtained first by using microscopy techniques.

It is worth pointing out at this stage that, although the equations for the Young's moduli are equally applicable in tension and in compression, in principle up to the failure strain of the composite, in practice this is seldom the case in compression. Fibre buckling can occur at the very early stages of deformation, especially in low volume fraction composites with relatively pliant matrices (Dow & Rosen, 1965).

Strength and fracture

Strength and fracture properties of fibrous composites can be approached in several ways which should be considered complementary rather than mutually exclusive. The complexity of the materials and the number of mechanisms which can initiate failure and fracture are such that no single body of theory can describe and predict accurately the initiation, evolution and final stages of the irreversible deformations associated with strength and fracture.

A brief outline of the various approaches is given in this section; it is necessarily brief and reference should be made to the relevant literature for a more comprehensive discussion. A distinction is also made between strength and fracture theories; they are obviously related but the methodologies needed for their description are sufficiently different to be dealt with separately.

The simplest theories of strength for fibrous composites are based on the rule of mixtures discussed in the previous section (Kelly, 1973). Assuming that fibres and matrix have the same failure strain, equation (1) can be extended to give the tensile strength of the composite in the fibre direction:

$$\sigma_\ell^* = \sigma_f^* V_f + \sigma_m^*(1 - V_f) \tag{10}$$

where σ^* is the tensile strength of composite (ℓ), fibres (f) and matrix (m). If the fibres and the matrix do not have the same failure strain, which is often the case, equation (10) needs to be modified (Hull, 1981) but its basic structure does not change and it will still show that the composite tensile strength is a linear function of the fibre tensile strength. Equation (10) needs also to be modified if the fibres are not infinitely long or if they cannot be assumed to be so. In this case equation (10) becomes:

$$\sigma_\ell^* = \sigma_f^* \left(1 - \frac{l}{2l_c}\right) V_f + \sigma_m^* (1 - V_f) \tag{11}$$

where ℓ_c is the 'critical fibre length' and ℓ is the actual fibre length. The critical fibre length takes into account the fact that in a short fibre composite the stress in the fibres builds up from zero at the two ends to the fibre strength value through shear stresses acting at the fibre–matrix interface as shown in Fig. 3 (Cox, 1952).

Equations (1) and (10) and their modifications have no equivalent when the composite is stressed in compression parallel to the fibre direction, or in in-plane shear or in the transverse direction (tension and compression). This is not altogether surprising because the micromechanisms associated with failure initiation are different in all these cases. Compression failure parallel to the fibre direction is triggered by fibre buckling which depends on the elastic rather than strength properties of fibres and matrix. Transverse strength in tension and compression is

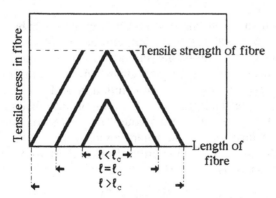

Fig. 3. Fibre stress as a function of fibre length (ℓ) in a short-fibre composite; definition of critical transfer length, ℓ_c.

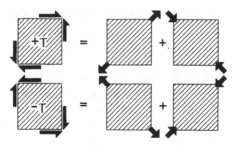

Fig. 4. Positive (+T) and negative (−T) shear stresses in a fibrous material expressed as a combination of tension and compression.

dominated by matrix properties. The shear strength can be thought of as strength under a combined state of stress, one tensile and one compressive of the same magnitude (Fig. 4); one important consequence of this is that the shear strength of anisotropic materials depends on the sign of the shear stress which is not the case for isotropic solids (Ashton, Halpin & Petit, 1969).

Of the five independent strength parameters needed to characterise even a simple unidirectional composite with parallel fibres, only one can be predicted with some confidence from the strength properties of the constituent phases. The others must be determined experimentally. It is also important to note that four out of five strength parameters are matrix-controlled, the only fibre-controlled one being the tensile strength parallel to the fibre direction. This has important implications when complex states of stress are present in these materials, as in the case of fibrous foods during biting and mastication, because the materials can fail almost simultaneously in more than one way making it difficult to identify the precise mode of fracture.

A second approach to the strength of fibrous composites is based on failure criteria. One accepts that the five strength parameters cannot be predicted but so long as they can be measured they can be incorporated into a macromechanical criterion for strength which does not concern itself with the microevents associated with initiation and evolution of damage but only with the final outcome. One important advantage of failure criteria is that they can take into account the interaction between failure modes, particularly important when complex stress states are present. All strength criteria can be expressed in compact form by the following equation (Tsai & Hahn, 1980):

$$F_i \, \sigma_i + F_{ij} \, \sigma_i \, \sigma_j = 1 \qquad (i,j = 1, 2, 3) \qquad (12)$$

where F_i and F_{ij} are strength related parameters and σ_{ij} are the applied stresses, three in the simple case of thin composites, two normal stresses and one in-plane shear (a repeated index implies summation over that index).

This kind of failure criterion has been successfully applied to the strength predictions of artificial fibrous composites. Its extension to more complex materials such as fibrous foods should be possible so long as the strength parameters have been measured and the effective state of stress has been determined.

The most severe limitation of failure criteria is that they cannot deal with cracks and crack propagation. This becomes the province of fracture mechanics. In its simple form it deals with the stability of cracks in stressed bodies by taking into account the stress concentration which arises near a crack tip:

$$\sigma = \frac{K}{\sqrt{(2\pi a)}} \tag{13}$$

where σ is the remote applied stress, a the crack length and K a parameter which measures the stress intensity at the crack tip. As σ increases for a given crack length a, K reaches a critical value K_c when the crack propagates in an unstable manner. This works well with brittle or semi-brittle materials where the crack length can be defined fairly precisely. Equation (13) has also been modified in a number of ways to allow for plastic flow at the crack tip. Its application to heterogeneous fibrous composites is debatable because in these materials there is a multitude of microevents associated with fracture which make it difficult to define with any accuracy an initial crack length. Also, crack propagation can develop simultaneously in more than one direction and self-similar crack growth is an implied assumption in equation (13) (Kanninen & Popelar, 1985, pp. 392–436).

A more promising approach to the fracture properties of heterogeneous, anisotropic materials is based on what is known as the 'energy method' (Gurney & Ngan, 1971). This approach is closer to the traditional Griffiths criterion for crack propagation (Griffiths, 1920) and both have the virtue of including an energy dissipation term which can be related to microevents associated with crack initiation and crack propagation (Atkins & Mai, 1985). In its simplest form, the energy method can be written as:

$$X\mathrm{d}u = \mathrm{d}E + R\mathrm{d}A \tag{14}$$

where X is an applied force, $\mathrm{d}u$ a corresponding displacement of the cracked body, $\mathrm{d}E$ the elastic energy stored in the body, $\mathrm{d}A$ the increment of cracked area and R the work dissipated irreversibly to increase the crack area by dA. With this method the fracture energies related to individual micromechanical events during damage evolution are all incorporated into R. In some instances they can be evaluated independently, either experimentally or theoretically, and their relative importance quantified. In fibrous composites the events associated with microdamage typically include fibre fracture (brittle or ductile), matrix cracking, matrix yielding, fibre–matrix debonding, fibre pull-out, etc.

At this stage of the development of food mechanics it is not possible to suggest a particular method as opposed to another for the study of strength and fracture of food materials. The energy method has been used with some measure of success in meat (Dobraszcyk et al., 1987), cheese (Luyten, 1988) and apples (Vincent et al., 1991). The results suggest that properly measured fracture properties can be related to textural attributes. Some work has also been done in relating fracture properties of plant materials to their characteristics as foods (Vincent, 1990).

Cellular materials

Cellular materials constitute a very important class of foods. Those of natural origin include all soft fruits and vegetables; among the manufactured types bread and associated products are widespread across the whole world. More and more cellular 'snack-type' products are being manufactured from a variety of raw materials. In spite of their importance, it is only in recent years that a rational approach to the mechanics of cellular structures has emerged.

Special textural attributes are associated with the progressive collapse of cell walls in air-filled food materials and with the release of cell contents from the liquid-filled structures of fruit and vegetables. In order to preserve, enhance or 'design-in' these desirable features it is important to understand the mechanics of deformation and fracture of these materials. Most of the work done in recent years has followed some earlier studies of polymeric foams, extending the range of applicability of the results and establishing a firm theoretical foundation for analysis (Hilyard, 1982; Gibson et al., 1982; Gibson & Ashby, 1988).

Elastic properties

A crucial property in the study of cellular materials is the ratio of densities between the cellular and the solid phase of the same substance. This provides not only a measure of the heterogeneity of the cellular material but also the link between the modes of deformation and fracture at the

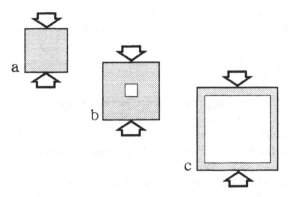

Fig. 5. The transition from a solid to a cellular material (see the text for explanation).

level of a single cell and the macroscopic bulk behaviour of the cellular structure.

Consider for example the three structures shown in Fig. 5 where the total volume of solid material is the same. If a compressive load is applied to material a, the deformation in the solid will be uniform. In material b the bulk of the solid will still experience a uniform state of strain, except for a small volume around the circular hole. Essentially b is a slightly perturbed equivalent of a. In 'material' c, however, the distribution of the solid material is such as to introduce local modes of deformation different from either a or b and its response to the applied load will be significantly different from the first two.

Material b is a high density structure, whereas material c is a low density one. The dividing line between high and low density is somewhat arbitrary but cellular systems with volume fractions of solids of 80% or more are considered to be high density, those with solids fractions of less than 10% are low density. Those in between can be considered to be either high or low density depending on available deformation modes.

The elastic properties of high density cellular solids can be obtained from rules of mixture, assuming that the gas inclusion has negligible properties compared to the solid. For the Young's modulus the following expression is obtained (Jeronimidis, 1988):

$$E_c = E_s \left(\frac{\varrho_c}{\varrho_s} \right) \tag{15}$$

where E_s is the Young's modulus of the solid, ϱ_s its density and ϱ_c the density of the cellular structure.

If bending deformations are also included, as will happen in a low density material, the expression for the modulus becomes (Ashby, 1983):

$$E_c = kE_s \left(\frac{\varrho_c}{\varrho_s} \right) \tag{16}$$

where K is a constant. The major difference between equations (15) and (16) is the different dependence on relative density.

If the cellular structure contains an incompressible liquid which cannot flow from cell to cell (closed cell structure), the interactions between cell walls and liquid must also be considered because the state of stress of the cell walls, and hence their possible modes of deformation are modified by the presence of the liquid under pressure. In effect the liquid under pressure stresses the cell walls in tension, preventing or delaying certain modes of deformation. This is the situation which exists in fruits and vegetables with turgor. The same considerations apply to interconnecting liquid-filled cellular structures where the rate of liquid flow from cell to cell is small owing to viscous and channel size effects (Warner & Edwards, 1988).

The various models presented in this section have been verified for a number of natural and artificial cellular materials. Their application to food structures has also been verified in a simple case not involving liquid-filled systems (Attenburrow et al., 1989).

Strength and fracture

In relation to textural attributes of cellular foods, their compressive strength and fracture properties are perhaps more important than their elastic behaviour. In this case too the relative density of the cellular material provides a link between strength and fracture properties of base substances and cellular structures. Depending on the properties of the solids from which the cellular materials are derived, three modes of failure of the cell struts (open cells) or cell walls (closed cells) are possible: elastic buckling, plastic collapse and brittle collapse. Elastic buckling occurs when the cell struts or walls in compression can undergo significant bending deformations before reaching their elastic strain limit. This depends both on the geometry of the cross-section of the wall or strut and on the mechanical properties of the solid. Plastic collapse follows elastic collapse for ductile materials, which can yield and deform plastically beyond their elastic limit. Brittle collapse is associated with solids with small strains to failure so that brittle fracture occurs before buckling. The expressions for the compressive strength of cellular materials for the three cases are (Gibson & Ashby, 1988):

$$\sigma^* = k_1 E_s \left(\frac{\varrho_c}{\varrho_s} \right)^2 \text{ (elastic collapse, open cells)} \tag{17}$$

$$\sigma^* = k_2 E_s \left(\frac{\varrho_c}{\varrho_s} \right)^3 \text{ (elastic collapse, closed cells)} \tag{18}$$

$$\sigma^* = k_3 \sigma_{sp} \left(\frac{\varrho_c}{\varrho_s} \right)^{3/2} \text{ (plastic collapse)} \tag{19}$$

$$\sigma^* = k_4 \sigma_{sb} \left(\frac{\varrho_c}{\varrho_s} \right)^{3/2} \text{ (brittle collapse)} \tag{20}$$

In these equations, σ^* is the compressive strength of the cellular structure, ϱ_c its density, E_s the Young's modulus of the solid, ϱ_s its density, σ_{sp} the yield strength of the solid and σ_{sb} its brittle strength; $k_1, \ldots k_4$ are constants.

The above expressions can be used to determine the maximum stress that a cellular structure can carry before collapsing reversibly, progressively or irreversibly depending on the properties of the solid. Obviously, if the properties of the solid change because of temperature, strain rate or water absorption effects, these changes can be transmitted to the cellular structure. Such changes are particularly significant when the behaviour of the solid changes from brittle (crisp texture) to plastic or rubber-elastic (soggy texture) as a result of water pick-up.

Also important from the point of view of perceived texture of cellular foods are the rules which govern the propagation of damage initiated by the mechanisms discussed above. Texture is likely to be affected by damage stability considerations. It is perhaps possible to differentiate quantitatively between a crisp and a crunchy texture, associating the former with unstable propagation of brittle collapse and the latter with stable, progressive propagation.

At present little is known about these aspects but they are perhaps among the most relevant ones for a better understanding of texture as a sequence of damage initiation, development and propagation.

Conclusions

The application to food materials of experimental and theoretical methods of investigation such as those presented here is just beginning. Progress is slow because of the complexities involved in characterisation, testing and analysis, but the available methods developed for other materials must in principle be applicable to food structures as well. The design and introduction of textural attributes in manufactured foods, in particular, can benefit greatly from these techniques. Further work is needed to establish food mechanics as a rational and science-based discipline.

References

Ashby, M.F. (1983). The mechanical properties of cellular solids. *Metallurgical Transactions*, **14A**, 1755–69.

Ashton, J.E., Halpin, J.C. & Petit, P.H. (1969). *Primer on Composite Materials: Analysis*. Technomic Publishing Co., Westport, CT.

Atkins, A.G. & Mai, Y.-M. (1985). *Elastic and Plastic Fracture*. Horwood, Chichester.

Attenburrow, G.E., Goodband, R.M., Taylor, L.J. & Lillford, P.J. (1989). Structure, mechanics and texture of a food sponge. *Journal of Cereal Science*, **9**, 61–9.

Cottrell, A.H. (1964). *The Mechanical Properties of Matter*. John Wiley & Sons, New York, London and Sydney.

Cox, H.L. (1952). The elasticity and strength of paper and other fibrous materials. *British Journal of Applied Physics*, **2**, 72–9.

Dobraszcyk, B.J., Atkins, A.G., Jeronimidis, G. & Purslow, P.P. (1987). Fracture toughness of frozen meat. *Meat Science*, **21**, 25–49.

Dow, N.F. & Rosen, B.W. (1965). Evaluation of filament-reinforced composites for aerospace structural applications. *NASA CR-207*, April.

Fellers, C. & Carlsson, L. (1979). Measuring the pure bending properties of paper. A new method. *Tappi Journal*, **62**, 107–9.

Gibson, L.J. & Ashby, M.F. (1988). *Cellular Solids – Structure and Properties*. Pergamon Press, Oxford.

Gibson, L.J., Ashby, M.F., Shajer, G.S. & Robinson, C.I. (1982). The mechanics of two-dimensional cellular materials. *Proceedings of the Royal Society of London*, Ser. A, **382**, 25–42.

Griffiths, A.A. (1920). The phenomenon of rupture and flow in solids. *Philosophical Transactions of the Royal Society of London*, Ser. A, **221**, 163–98.

Gurney, C. & Ngan, K.M. (1971). Quasistatic crack propagation in non-linear structures. *Proceedings of the Royal Society of London*, Ser. A, **325**, 207–31.

Hashin, Z. (1962). The elastic moduli of heterogeneous materials. *Journal of Applied Mechanics*, **17**, 143–50.

Hilyard, N.C. (ed.) (1982). *Mechanics of Cellular Plastics*. Applied Science, London.

Hull, D. (1981). *An Introduction of Composite Materials*. Cambridge University Press, Cambridge.

Jeronimidis, G. (1988). Structure and properties of liquid and solid foams. In *Food: Its Creation and Evaluation*, ed. J.M.V. Blanchard & J.R. Mitchell, pp. 59–74. Butterworths, London.

Jeronimidis, G. & Vincent, J.F.V. (1984). Composite materials. In *Connective Tissue Matrix*, ed. D.W.L. Hukins, pp. 187–210. Macmillan, London and Basingstoke.

Jones, R.M. (1975). *Mechanics of Composite Materials*. McGraw-Hill, Tokyo.

Kanninen, M.F. & Popelar, C.H. (1985). *Advanced Fracture Mechanics*, Oxford University Press, Oxford.

Kelly, A. (1973). *Strong Solids*. Clarendon Press, Oxford.

Khan, A.A. (1989). Mechanical properties of fruit and vegetables. Ph.D. thesis, University of Reading.

Lillford, P.J. (1988). The polymer/water relationship – its importance for food structure. In *Food: Its Creation and Evaluation*, ed. J.M.V. Blanchard & J.R. Mitchell, pp. 75–92. Butterworths, London.

Luyten, H. (1988). The rheological and fracture properties of gouda cheese. Ph.D. thesis, The Agricultural University, Wageningen.

Mai, Y.-M. & Atkins, A.G. (1980). Crack stability in fracture toughness testing. *Journal of Strain Analysis*, **15**, 63–74.

Tsai, S.W. (1987). *Composites Design*. Think Composites, Dayton.

Tsai, S.W. & Hahn, H.T. (1980). *Introduction of Composite Materials*. Technomic Publishing Co., Westport, CT.

Vincent, J.F.V. (1990). Fracture properties of plants. *Advances in Botanical Research*, **17**, 235–87.

Vincent, J.F.V., Jeronimidis, G., Khan, A.A. & Luyten, H. (1991). The wedge fracture test – a new method for measurement of food texture. *Journal of Texture Studies* **22**, 45–57.

Warner, M. & Edwards, S.F. (1988). A scaling approach to elasticity and flow in solid foams. *Europhysics Letters*, **5**, 623–8.

Williams, A.A. & Atkin, R.K. (eds.) (1983). *Sensory Quality in Foods and Beverages: Definition, Measurement and Control*. Horwood, Chichester.

J. F. V. VINCENT

Texture of plants and fruits

This chapter is about the problems that animals have when they try to feed on plants. Some bits of plants (e.g. fruits) are adapted to be eaten, but most of the rest of plants is not. The act of feeding on a plant involves fracture at a number of size levels. The possibility that parts of plants may have mechanisms to resist fracture by animals feeding on them has not been seriously and competently considered, except possibly for a small range of plants associated with domestic animals. In this instance, plants have been developed which are more 'palatable'. Since palatability is mostly measured in terms of intake, this often means that the plant is easier to harvest (e.g. easier to grasp, as Owen (1976) suggests for geese harvesting grasses) or weaker (Theron & Booysen, 1966; Owen, Nugent & Davies, 1977), though if the plant is too weak or brittle then the plant can break down too quickly inside the animal, which can lead to bloat. However, brittleness could be a means of defence by the plant, i.e. by abscission, which limits damage in the main part of the plant when the animal harvests part of it, or by limiting the amount of nutrient available to the animal. A brittle fracture would go straight through an assemblage of cells, opening as few cells as possible and thus releasing as little nutrient as possible: a toughening mechanism would inevitably spread damage within the plant material, causing more cells to be damaged, releasing more nutrient. An animal might therefore learn to reject a brittle plant which released little food. Such a mechanism would be a defence only against animals which chewed their food only once, such as plant-eating cyprinids (Sibbing, this volume). However, the role of toughness is not totally clear: leaf-cutting ants were reported to prefer less tough leaves (Waller, 1982), but in this study not only was toughness measured by the highly questionable technique of measuring the force required to push a pin through the leaf, but the tougher leaves were also older and may well have been chemically different. A wilted leaf, at lower turgor, is tougher (Vincent, 1990) and is probably less likely to have changed chemically during wilting. Even so, the grasshopper *Melanoplus*

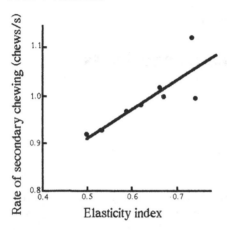

Fig. 1. An 'elasticity index', typical of the *ad hoc* measurements of mechanical 'parameters' of plant material. A sample of grass is placed in a cylinder and compressed by a piston. After a number of cycles of the piston up and down, the sample is compressed, then the piston retracted until no force is registered. In most cases the piston will not have returned to its original position at this point, allowing estimation of permanent or plastic deformation of the sample. The ratio between plastic (i.e. irrecoverable) deformation to total deformation is the 'elasticity index'. Here it seems to correlate with the rate of chewing during rumination. (From unpublished results of M.N. McLeod and D.J. Minson.)

differentialis prefers wilted grass (Lewis, 1982). The voluntary intake by cattle of some grasses and legumes was related to various physical and chemical attributes (M.N. McLeod & D.J. Minson, unpublished data). Several factors are correlated with reduced intake: 'grinding energy' (energy required to grind a dried sample of the plant to a given particle size) and the proportions of cellulose and lignin. In a more recent test (D.J. Minson, unpublished data), wet forage was ground between two quasi-concentric cylinders with ribs running at an angle to each other. This produced particles indistinguishable from those chewed by a cow and collected through a fistula and confirmed that 'grinding energy' is correlated with reduced intake. The rate of chewing during rumination is related to the 'elasticity index', a rather unsatisfactory parameter derived by compressing a sample of randomly arranged leaf and stem in a cylinder and noting the percentage elastic recovery when the force has been cycled back to zero (Fig. 1). It is probably inversely related to plasticity and may possibly be related to crispness.

Acceptability of plant material and rate of chewing in grass carp seem to be unrelated to mechanical properties. Carp took less *Lemna minor* than the grass *Glyceria* sp., but chewed both at the same rate. *Elodea canadensis*, much more brittle and less fibrous than grass, was taken at about the same rate as grass. However, the rate of intake (both in terms of speed and duration) is certainly very dependent on mechanical properties (Vincent & Sibbing, 1992).

Obviously many plants, especially xerophytes, have spiney or sharp leaves or fruits, which make them painful, if not dangerous, to eat. Such defences are not present where there are few or no herbivores, as on Hawaii (Lamoureaux, 1983). Otherwise, most plants rely on biochemical defences and the capacity of the animals feeding on them to learn what is safe to eat and what is not. This may imply that it is energetically cheaper for a plant to defend itself with a potent allelochemical, which will have to be made in only small quantities if it interferes with a ubiquitous metabolic pathway, than to lay down some mechanical defence. In addition, since the morphological level at which the plant is fractured depends largely on the size of the animal involved, it may well be a waste of time and effort for the plant to strengthen or toughen itself, since it may well be defending itself against damage from a particular size range of herbivore. The biochemistry of herbivores is probably much more predictable and general than the way they go about feeding.

Some plants respond to being eaten by proliferating. This has led to the proposition that it may even be beneficial for a crop plant to suffer limited attack. The kelp, *Egregia laevigata*, responds to damage by the grazing limpet, *Acmaea insessa*, by producing more branch rachises (Black, 1976). It also seems possible that reduction of the size of the kelp by grazing may even be beneficial, keeping it too small to be detached and swept away by the stormy sea.

Cells and tissues

Cellulose is one of the more stable biological materials and is therefore rather more difficult to digest. The contents of plant cells are much more nutritious than their containers, so it makes sense for the animal to go for those parts of the plant which have the lowest cellulose:cell contents ratio, which will also be the easiest to harvest. This is more or less the way fruits are designed and also applies to the younger parts of plants. The goal of the herbivore is probably to reach the contents of as many cells as possible with the least expenditure of energy. Ultimately cells have to be opened. This is a factor many people seem to have overlooked: most mechanical tests on plant tissues have been performed at the tissue level (e.g. plant fibres, wood, storage parenchyma) but few tests have differentiated between fracture at the tissue and at cellular levels. The differentiation is very important for an animal trying to reach the contents of cells. Something is known, more or less anecdotally, of the way in which bacterial action in the rumen of cattle can weaken cell walls, but attempts to put numbers to the mechanical properties involved have been less than satisfactory. For instance Evans, Burnett & Bines (1974) correlated the strength of some monocotyledonous leaves with the fibre content and showed that it was reduced when the leaves had been incubated in the rumen. They were measuring the mechanical properties of tissues rather than of cell walls. Even so, the time required for the strength of the leaves to drop to half its original value when they were incubated in the rumen correlated well with the total time taken by the cow for eating and ruminating the same leaves.

It is possible to correlate the tendency of plants to cause bloat in ruminants (due to the cells breaking open too easily, releasing their contents all at once) with the mechanical strength of the cells and tissues (Lees *et al.*, 1981, 1982; Lees, 1984). Lees took leaf tissue of a number of forage legumes, some of which caused bloat and some of which did not, and crushed them with glass beads, ground them in an homogeniser or broke them open by sonication. In each instance the number of cells broken open was estimated from the amount of chlorophyll released. Legumes which cause bloat have weak cells; the cell walls are thinner and the cells are more readily broken open (Fig. 2); those which do not cause bloat have stronger and thicker cell walls, but the tissue can be strong or

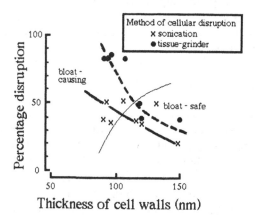

Fig. 2. Cells with thicker cell walls are more difficult to break and less likely to cause bloat in cattle eating them. (Drawn from data tabulated in Lees, 1984.)

weak. But there is no correlation between the thickness of parenchyma cell walls and the strength of the tissue (e.g. fibre). This may well appear to be obvious when pointed out, but it shows clearly the necessity of considering fracture at a number of levels of organisational hierarchy: something which is rarely done with plants. The stronger tissues, as in sainfoin, have more complex and more reticulate venation, the veins frequently being attached to both layers of epidermis by bundle sheath cells, which divides the mesophyll tissue into compartments and prevents loss of cells. In birdsfoot trefoil the epidermal layers are not connected in the same way, so once the leaf has been ruptured, mesophyll cells can be forced from between the epidermal layers and be broken down more easily. However, when the effects of bacterial breakdown, as in ruminants, are considered, the degree to which cells are stuck together is also important. At this point in the process, the compartmentalisation of the leaf lamina by the reticulate arrangement of veins in a dicotyledonous leaf confines digestion to discrete compartments. The firm attachment of both epidermal layers to bundle sheath cells and their extensions and the low digestibility of these two components serves to restrict digestion within the leaf. Thus milk vetch shows such compartmentalisation and is bloat-safe, whilst alfalfa does not and will cause bloat. An interesting point here is that the digestibility of a leaf can be affected by its morphology whilst not affecting any mechanical parameters. Simply by changing the degree of reticulation of the leaf both fracture and microbial breakdown can be changed.

Another aspect of the mechanical properties of plant tissue is 'crispness', which is probably a form of brittleness. It is not at all clear where crispness originates; perhaps it depends on the level at which fracture occurs whether one considers it to arise from properties of cell or tissue. Since it is a brittle fracture it will occur at high turgor and invoke few if any of the resident toughening mechanisms, suggesting that it depends on good shear connection between cells. The reason why chilled plants are crisper is not clear. There is a number of possibilities: low temperature could reduce the viscous component in the pectins of the primary lamina, giving better shear transfer of stored elastic (strain) energy to the advancing crack; or the cellulose, being viscoelastic, could increase in stiffness, leading it to achieve a smaller strain at a given turgor stress. This in turn would increase the turgor pressure by reducing the volume of the cell, leading to higher stresses in the cell walls, to better shear transfer of elastic strain energy, and bring the cell wall closer to its failure stress.

Leaves of most dicotyledonous plants, with their network of veins, can deflect cracks; the parallel arrangement of veins in the leaves of monocotyledonous plants has a rather different effect. The leaves of such

plants can be modelled as a composite material with preferentially orientated fibres (Vincent, 1982). Grass leaves with a low (less than 10% (u/v))
content of sclerenchyma and other fibres are almost completely notch-
insensitive (Fig. 3a). This is most probably because the shear stiffness of
the cells between the fibres is relatively low. Thus even if several fibres
have been broken, stress is not transmitted laterally sufficiently to cause a
stress concentration in the remaining fibre(s) and a notch will not weaken
the leaf significantly. This has a number of important consequences for
animals feeding on it, since it means that teeth are of little use other than
for gripping the grass, unless they work with a scissors action like the
incisors of rodents. Large animals such as cows and sheep hold the grass
(respectively with tongue and teeth) and pull. They thus have a limitation
on the number of leaves which they can break, illustrated by considering
the effects of 'enrichment' of pastures. Longer grasses will be harvested
at least as easily as shorter ones, so it is worth encouraging grasses to
grow tall. But if the grass tillers more densely the cow or sheep, having
finite strength, will find the size of its bite reduced, even though it might
be taking in the same number of grass leaves. This is because the strength
of grass and its apparent stiffness (as measured from the total cross-
section area of the leaf) are directly proportional to the amount of fibre
present. Thus increasing tillering will increase intake only up to a point
(A. Antuna, personal communication). Another example of the interaction between animals and their food plants is given by deer. When deer
age, their teeth tend to rot. If the back teeth remain, food intake is not
impaired, although the deer cannot then feed on shorter, younger and
more nutritious grass. However if the back teeth decay the deer cannot
survive, even if their front teeth are in good condition. Clearly deer do
not need their front teeth for gathering grass although they clearly aid
selectivity, but the back teeth are needed for comminution. An exception
to the apparent general rule that large animals do not cut grass when they
harvest it seems to be given by the kangaroo, whose front teeth seem to
be adapted for cutting. The front teeth of the camel appear to be similar.
This seems to be associated with the ability of these animals to select
small amounts of green vegetation at the base of large clumps of dry
plants (G. Sanson, personal communication). Thus the kangaroo can
exist by highly selective grazing in areas where less-selective sheep cannot
survive. This strategy also inflicts far less general damage on the vegetation. Grasses with higher fibre contents (greater than 15% by volume)
such as Deschampsia flexuosa and Stipa gigantea, whilst much stronger,
are sensitive to damage (Fig. 3b), presumably because the fibrous
material is continuous across the leaf just beneath the epidermis, thus
allowing the mechanisms of stress concentration to drive the fracture

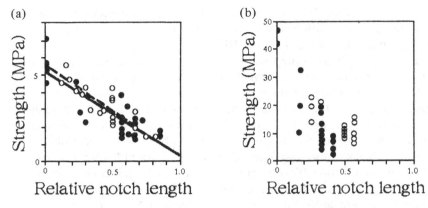

Fig. 3. Notch tests on grass leaves. (a) A notch will weaken material by reducing the cross-sectional area over which load can be supported in tension, in which case the strength will drop in direct proportion to the relative width of the notch (e.g. notch-sensitivity of *Lolium perenne*). (b) The notch can also weaken due to concentration of stress at the tip of the notch, when strength will drop off faster than in simple proportion to the length of the notch (e.g. notch-sensitivity of *Stipa gigantea*). Open circles in (b) and dashed line in (a) refer to edge notches, closed circles in (b) and solid line in (a) to centre notches.

(Vincent, 1991). These grasses will appear more brittle or 'dry', which may make them more susceptible to damage by tramping, although the large amount of fibre will make them either impossible to harvest (due to their greatly increased strength) or unpalatable. The epidermal cells have been implicated as a major load-bearing component in *Lolium perenne* (Greenberg *et al.*, 1989), although the implications of this idea on notch-sensitivity were not explored.

The animal has still more problems when it has broken a piece of leaf away. Many animals (e.g. fish, birds) do not break the plants down much further and extract the contents only of the cells around the edges of the broken bits. This leads to the observation that the smaller the piece of plant tissue the greater the amount of cellular damage. This is not quite so simple as saying 'the more the food is chewed the smaller it gets', since most of the damage is still around the edges (J.F.V. Vincent & F.A. Sibbing, unpublished results). Obviously it is preferable to break cells open by chewing them since then cells in the middle of a piece can be used. For some reason many herbivores don't seem to manage this and vast amounts of plant material go through their guts more or less intact. It is possible to get some idea about fracture at the cellular level from a simple sum. The toughness of parenchyma is of the order of 300 J m^{-2}

(Atkins & Vincent, 1984; Khan, 1989). If the average diameter of a cell is 0.1 mm then there will be about 10^8 cells per square meter, giving a work to fracture of a single cell of 3×10^{-6} J.

Fracture mechanics is dependent not only on material and structure but also on size. In general, smaller objects are tougher, since there is relatively less volume (a term in length cubed) for the storage of strain energy to feed to the advancing fracture (whose area is a term in length squared). Also, since the length of a critical crack remains the same, being a property of the material, a smaller object will be less likely to contain a crack of this length and therefore be safer at the same loads. It therefore seems possible that cells are protected against being broken open simply by their small size. Given the stiffness of cellulose as 10 GPa, work of fracture as 10^3 J m^{-2}, turgor pressure of 1 MPa, cell diameter of 10 μm and cell wall thickness of 0.1 μm, the Griffith critical crack length (see Jeronimidis, this volume; Gordon, 1968) is 25 mm! Since this length is proportional to the work of fracture, the latter quantity would have to be reduced to that of a glass (1 J m^{-2}) before the crack length approached the dimensions of a cell. Alternatively the internal pressure would have to be increased to 10 MPa. This is close to the force (76 kg cm^{-2}, i.e. about 8 MPa) quoted by Sanson (Anon., 1988) necessary to fracture plant material in the mouth. This increase in internal pressure would increase the strain in the cell wall from 10^{-3} to 10^{-2}, implying that the deformation of the cell required to make it fracture in a brittle manner is in reality quite small. The problem for the animal is much more likely to be positioning the cell sufficiently accurately between two apposed surfaces to be able to compress it. This suggests that the more unhomogeneous (e.g. fibrous) the plant material is, the easier it is to break down by chewing. Green Panic (*Panicum maximum*) containing 1.7% volume fraction of sclerenchyma broke down much more completely during rumination than did Italian ryegrass (*Lolium multiflorum*) containing only 0.7% volume fraction of sclerenchyma (Wilson *et al.*, 1989). Similarly Koegel, Fomin & Bruhn (1973), who studied methods of removing cell fluid from alfalfa by squeezing it between rollers, found that the more fibre there was in the plant the more fluid could be extracted. Comparing these results with those of Lees *et al.*, one gets the impression that it is far easier to break cells when they are separate individuals than when they are incorporated into an unhomogeneous tissue, but cells in mechanically uniform tissues seem to be well protected individually. The presence of fibre can protect the tissue from fracture but may, during chewing, supplement the action of the teeth by concentrating stresses at the cellular level. Bacterial action in ruminants is probably most important to separate cells from each other. This correlates with the idea that the main factor affecting transport of material from the rumen in cows is particle size (D.J. Minson, personal communication) such that particles have to be 1 mm or less in their major dimension before they can leave the rumen. This 1 mm criterion probably indicates that all breakdown necessary for further digestion has occurred, since during rumination it is far easier to break the leaf particles between the fibres than across them (Wilson, McLeod & Minson, 1989).

Scale effects; seeds

Since fracture mechanics are dependent on size as well as material and structure (see above), smaller plants need adapt far less for toughness and structural integrity. The evolutionary increase in the size of plants must be related not only to the development of supportive structures but also to the development of suitable toughening mechanisms, although the size ranges over which the available toughening mechanisms become available or are required has not been determined. Perhaps all that is important for the plant is the limiting smallness where, for a particular structure made of a particular material, fracture becomes so difficult or rare that a particular size confers a significant selective advantage. This principle will also apply to parts of a plant, for instance seeds.

A small seed will be more difficult to break open and can therefore more readily survive being eaten. Thus a fruit can have either a few large seeds which are commonly protected by a hard shell, or a large number of small seeds which need not have such extreme protection and may even have large amounts of edible material associated with them which will encourage the attentions of an animal. This implies that the seeds are as well protected from the trituration mechanisms of the animal as they are from the more commonly assessed digestive chemical mechanisms. Also it is probably quite difficult to get a small seed in between the teeth where it can be chewed. There are several interesting questions here which are considered by Lucas (this volume). What sizes are nuts and naked seeds? Do the size ranges overlap? How does the thickness of the nutshell (if present) scale with size? Does a nut inside a fruit, which is presumably 'meant' to be eaten, have a different sort of shell from (say) an acorn? Thickness or composition? The size of grit in the gizzard of seed-eating birds seems to be more related to the size of seed preferred than to the size of the bird or the size of grit most readily available (M. Owen, personal communication). What is the degree of mechanical damage to seeds which have been eaten? How many survive this trauma? Remarkably, there is some information in this area. Paulsen (1978) measured some mechanical properties of soybeans in compression. He recorded the strain energy density (which he mistakenly called 'toughness') at the load when the seedcoat ruptured. There is a very marked effect of size such that the smaller (5.95 to 6.35 mm long) soybeans store up to twice the energy per unit volume before fracture than do the larger (6.75 to 7.14 mm long) soybeans.

Fruits and skins

The parenchyma of most fruits is relatively 'tender' as might be expected for something which has evolved to be eaten by animals. The measured work of fracture is between 100 (radial) and 300 (tangential) J m^{-2} for a normal, mid-season 'eating' apple such as a 'Cox' (Khan, 1989). The significance of the orientation is that the cells are arranged in radial columns (Reeve, 1953; Vincent, 1989) which direct the path of a free-running crack, encouraging it to travel radially (Khan, 1989). The toughness can go as high as 1 kJ m^{-2} with very late, dense, apples such as the 'Norfolk Beefing' (Khan, 1989). In general it seems that the mechanical and fracture properties of apple parenchyma are controlled by a combination of cellular orientation and density: late maturing apples tend to have a greater proportion of orientated cells and to be more dense, hence more anisotropic and more stiff. The more anisotropic apples tend to be more brittle when fractured radially (J.F.V. Vincent & A.A. Khan, unpublished data).

The fracture properties of fruit skin are obviously of importance, since the skin provides protection from mechanical damage, e.g. by teeth. There is no evidence that fruit skin is anisotropic in such a way as to resist fracture. The lenticels in the skin could act as starter cracks (Brown & Considine, 1982), in which case they would be expected to be orientated normal to the most probable direction of fracture and to be shorter than the critical (Griffith) length. There is some evidence for this in the plum where the shape of the fruit (cylindrical) and the strains observed in the intact fruit indicate that a vertical (i.e. longitudinal) split is much more likely. However, most splits are horizontal at an angle which is within a fraction of a degree of the mean orientation of the long axis of the lenticels. The lenticels are reported to give a stress concentrating factor of 3 (Mrozek & Burkhardt, 1973). In the grape this effect has been avoided by reinforcement around the lenticels, although the reinforcement itself serves to concentrate stress, leading to the formation of micro-ring fractures around them (Considine, 1982).

Fruit skins may or may not be notch-sensitive, depending on variety. Amongst apples, 'Granny Smith' has a relatively notch-sensitive skin whereas 'Christmas Pearmain' is insensitive to damage but the work to

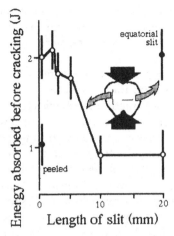

Fig. 4. The effect of skin on energy absorption by an apple ('Rock Pippin') compressed vertically to the core. The skin increases total energy absorbed unless it contains a vertical crack longer than a certain length (here, 10 cm), when it ceases to protect the apple and the apple behaves as if it had no skin at all. An equatorial crack does not affect energy absorption. (From Khan, 1989.)

fracture is very different for the skin of different fruits. Whereas apple skin has a fracture toughness of the order of 200 J m^{-2}, tomato skin can be less than 10 J m^{-2} (J. F. V. Vincent, unpublished observations). It may be unwise to compare experiments with isolated skin with the performance of the skin when it is on the fruit, since, especially in the apple where it is difficult at a histological level to say where the skin begins, the cells beneath the skin must be providing very significant support. Indeed, when a piece of 'isolated' apple skin fractures, the outer, more homogeneous 'cuticle' layer fractures well in advance of the two or three layers of cells which invariably stick to the inside of the skin, suggesting that the cells are retarding the propagation of the crack (J.F.V. Vincent, unpublished observation). When the skin on an intact ripe fruit does break, the internal structure may be providing a significant amount of the fracture energy. Conversely, the skin protects the inner tissues from damage by cracking. When a 'New Rock Pippin' apple has a vertical crack of longer than 5 mm (a sixth to an eighth of the height of the apple) in the skin, force and deformation at failure (by cracking rather than bruising) drop off very rapidly. When the crack is more than 10 mm long, the apple cracks as if it had no skin at all (Fig. 4; Khan, 1989). Although we have not measured it, I suspect that the important factor in fruit skin is not its resistance to crack propagation but rather its resistance to cracks starting. The skin may therefore be very much like many stiff commercial packaging plastic films of the type which prevent you from reaching the food so hygienically packed. The biggest problem the herbivore has is getting its teeth in to start with.

As regards the role of skin in protecting the fruit, the common mistake is to assume that the strongest skin will be the most resistant (see e.g. Considine, Williams & Brown, 1974). But fracture mechanics tell us that strength is not the main factor; the crack has to be initiated (which will be governed by the presence and orientation of stress-concentrating defects) and propagated (which will be governed by the transmission of energy to the advancing crack tip). The best protection will be gained from a tough skin which is therefore likely to be relatively extensible (Lustig & Bernstein, 1985). However, the one-dimensional strains observed in isolated skins, of the order of 15% to 25% are never achieved when the skin is on the fruit. This is partly because the skin is being supported by the cells beneath and partly because the skin on the fruit is being stretched in two directions at once and the Poisson ratio effects (i.e. the narrowing of the sample observed when it is stretched in a uniaxial test) are not available. The skin is then less deformable and its stiffness increases. The analysis of fracture in two-dimensional strain is very diffi- cult or even insoluble. Very often the best approach is experimental. True two-dimensional strain can be achieved in a number of ways such as mounting the skin over a hole as a diaphragm, then pressurising it and measuring pressure and deflection of the centre of the diaphragm (Voisey & Lyall, 1965). Unfortunately Voisey & Lyall did not include sufficient information in their paper to allow calculation of any fracture parameters. They did quote the pressure to cause fracture of the diaphragm and showed that this is not well correlated with susceptibility to cracking. This is not surprising, since the deflection of the diaphragm (which they did not measure) is also needed in order to calculate strength and would, if incorporated into the calculations, probably improve the correlation with cracking. Another technique is to pressurise the whole fruit and measure how much the skin stretches (Lustig & Bernstein, 1985). This is much more representative of the conditions with which the skin of a fruit has to cope and results in fracture at lower strains averaging 7%.

Nuts and endosperm

Some nuts are obviously very brittle, such as Brazil and hazel. No measurements have been published, although Lucas and co-workers are currently working in this area. The endosperm of *Zea mays* has a remarkably high toughness of over 1 kJ m^{-2} (Balastriere, Herum & Blaisdell, 1982; Mensah *et al.*, 1981; Vincent, 1990) but the vegetable ivory of the ivory nut palms must be tougher even than this. No measurements have been made.

Some final thoughts

As far as I know, no-one has developed any ideas on how size of animals interacts with plants at various levels of hierarchy or experimented in this area. The radular teeth of herbivorous molluscs are always related in size to the mechanical properties of the food rather than to any scaling factor of the mollusc itself (A. Graham, personal communication). This implies that molluscs feed on plants at only the cellular level: the radula only ever breaks open cells and is never used for breaking plants at the tissue level. A different type of interaction occurs between grass and grazers. Larger animals simply grip and pull (see above). Mid-size animals such as geese grip the grass then bend it sharply before pulling it quickly. The bend probably weakens the grass in the same way that a fold weakens a piece of paper. The quick pull suggests a strain rate effect, though the only data available suggest that toughness and strength *increase* with increasing strain rate (Greenberg *et al.*, 1989). Smaller animals still (rabbits, locusts etc.) have to cut through the individual fibres of the leaf, since they are not strong enough to break the grass in tension. These scale effects could be used to 'tune' the mechanical properties of plants to favour herbivores of a certain size range. Thus one could develop plants which would be palatable to and nutritious for large animals such as cattle, but not for smaller ones such as geese. This would open the way for benign control of their grazing, keeping them clear of growing crops but perhaps attracting them to another, non-sensitive, area where plants more suitable for the smaller bird are cultivated. Since plants seem to be fairly modular in their design and are made from a limited number of materials, a study of scale effects in the interactions between plants and herbivores would be of great interest and value.

References

Anon. (1988). Teeth: the cutting edge of ecology. *Monash Review*, August 1988, 6–7.

Atkins, A. G. & Vincent, J. F. V. (1984). An instrumented microtome for improved histological sections and the measurement of fracture toughness. *Journal of Materials Science Letters*, **3**, 310–13.

Balastreire, L. A., Herum, F. L. & Blaisdell, J. L. (1982). Fracture of corn endosperm in bending. Part II. Fracture analysis by fractography and optical microscopy. *Transactions of the American Society of Agricultural Engineers*, **25**, 1062–5.

Black, R. (1976). The effects of grazing by the limpet, *Acmaea Insessa*, on the kelp, *Egregia laevigata*, in the intertidal zone. *Ecology*, **57**, 265–77.

Brown, K. & Considine, J. (1982). Physical aspects of fruit growth: stress distribution around lenticels. *Plant Physiology*, **69**, 585–90.

Considine, J. A. (1982). Physical aspects of fruit growth: cuticular fracture and fracture patterns in relation to fruit structure in *Vitis vinifera*. *Journal of Horticultural Science*, **57**, 79–91.

Considine, J. A., Williams, J. F. & Brown, K. C. (1974). A model of studies on stress in dermal tissues of mature fruit of *Vitis vinifera*: criteria for producing fruit resistant to cracking. In *Mechanisms of Regulation of Plant Growth*, ed. R. L. Bielski, A. R. Ferguson & M. M. Cresswell, pp. 611–17. The Royal Society of New Zealand, Bulletin No.12.

Evans, E. W., Burnett, J. & Bines, J. A. (1974). A study of the effect of exposure in the reticulo-rumen of the cow on the strength of cotton, grass, hay and straw. *British Journal of Nutrition*, **31**, 273–84.

Gordon, J. E. (1968). *The New Science of Strong Materials*. Penguin, Harmondsworth, Middx.

Greenberg, A. R., Mehling, A., Lee, M. & Bock, J. H. (1989). Tensile behaviour of grass. *Journal of Materials Science*, **24**, 2549–54.

Khan, A. A. (1989). Mechanical and fracture properties of fruit and vegetables. Ph.D. thesis, University of Reading.

Koegel, R. G., Fomin, V. I. & Bruhn, H. D. (1973). Cell rupture properties of alfalfa. *Transactions of the American Society of Agricultural Engineers*, **16**, 712–16.

Lamoureaux, C. H. (1983). Plants. In *Atlas of Hawaii*, 2nd edn, ed. R. W. Armstrong, p. 72. University of Hawaii Press, Honolulu.

Lees, G. L. (1984). Cuticle and cell wall thickness: relation to mechanical strength of whole leaves and isolated cells from some forage legumes. *Crop Science*, **24**, 1077–81.

Lees, G. L., Howarth, R. E. & Goplen, B. P. (1982). Morphological characteristics of leaves from some forage legumes: relation to digestion and mechanical strength. *Canadian Journal of Botany*, **60**, 2126–32.

Lees, G. L., Howarth, R. E., Goplen, B. P. & Fesser, A. C. (1981). Mechanical disruption of leaf tissues and cells in some bloat-causing and bloat-safe forage legumes. *Crop Science*, **21**, 444–8.

Lewis, A. C. (1982). Leaf wilting alters a plant species ranking by the grasshopper *Melanoplus differentialis*. *Ecological Entomology*, **7**, 391–5.

Lustig, L. & Bernstein, Z. (1985). Determination of the mechanical properties of the grape berry skin by hydraulic measurements. *Scientia Horticulturae*, **25**, 279–85.

Mensah, J.K., Herum, F.L., Blaisdell, J.L. & Stevens, L.L. (1981). Effects of drying conditions on impact shear resistance of selected corn varieties. *Transactions of the American Society of Agricultural Engineers*, **24**, 1568–72.

Mrozek, R.F. & Burkhardt, T.H. (1973). Factors causing prune side cracking. *Transactions of the American Society of Agricultural Engineers*, **16**, 686–92.

Owen, M. (1976). The selection of winter food by whitefronted geese. *Journal of Applied Ecology*, **13**, 715–29.

Owen, M., Nugent, M. & Davies, N. (1977). Discrimination between grass species and nitrogen-fertilised vegetation by young Barnacle geese. *Wildfowl*, **228**, 21–6.

Paulsen, M.R. (1978). Fracture resistance of soybeans to compressive loading. *Transactions of the American Society of Agricultural Engineers*, **21**, 1210–16.

Reeve, R.M. (1953). Histological investigations of texture in apples. II. Structure and intercellular spaces. *Food Research*, **18**, 604–17.

Theron, E.P. & Booysen, P. de V. (1966). Palatability in grasses. *Proceedings of the Grassland Society of South Africa*, **1**, 111–20.

Vincent, J.F.V. (1982). The mechanical design of grass. *Journal of Materials Science*, **17**, 856–60.

Vincent, J.F.V. (1989). The relation between density and stiffness of apple flesh. *Journal of the Science of Food and Agriculture*, **47**, 443–62.

Vincent, J.F.V. (1990). Fracture properties of plant tissue. *Advances in Botanical Research*, **17**, 235–87.

Vincent, J.F.V. (1991). Mechanical strength and design of grasses. *Journal of Materials Science*, **26**, 1947–50.

Vincent, J.F.V. & Sibbing, F.A. (1992). How the grass carp (*Ctenopharyngodon idella*) chooses and chews its food: some clues. *Journal of Zoology* (in press).

Voisey, P.W. & Lyall, L.H. (1965). Methods of determining the strength of tomato skins in relation to fruit cracking. *Proceedings of the American Society for Horticultural Science*, **86**, 597–609.

Waller, D.A. (1982). Leaf-cutting ants and live oak: the role of leaf toughness in seasonal and intraspecific host choice. *Entomologia Experientia et Applicata*, **32**, 146–50.

Wilson, J.R., McLeod, M.N. & Minson, D.J. (1989). Particle size reduction of the leaves of a tropical and a temperate grass by cattle. I. Effect of chewing during eating and varying times of digestion. *Grass and Forage Science*, **44**, 53–63.

Wilson, J.R., Atkin, D.E., McLeod, M.N. & Minson, D.J. (1989). Particle size reduction of the leaves of a tropical and a temperate grass by cattle. II. Relation of anatomical structure to the process of leaf breakdown through chewing and digestion. *Grass and Forage Science*, **44**, 65–75.

P. P. PURSLOW

Measuring meat texture and understanding its structural basis

The flesh of animals, and especially the muscle tissue, is a highly nutritious food. There are many carnivores and omnivores that have evolved to hunt or scavenge and eat the flesh from other animals; however in this chapter I shall concentrate on the human consumption of meat in a modern context, and particularly on our understanding of its textural qualities. It is a reasonable assumption that, in prehistoric times, obtaining a sufficient supply was the overriding priority. When there was enough to eat, the wholesomeness of the meat (i.e. freedom from disease, parasites and bacteriological spoilage) was most probably the first aspect of choice. The practice of cooking of meat may have conceivably come about as a way of reducing the risk of eating infected or rotting meat. In some areas of the world today, being able to obtain or afford meat is still a problem for many people. In Western society the supply and the wholesomeness of the supply of meat is largely taken for granted (although recent worries about *Salmonella*, *Listeria* and bovine spongiform encephalopathy show that wholesomeness is still an overriding concern) and the eating quality of meat – its taste, juiciness and texture – becomes a great consideration.

Food texture is an attribute that is determined principally in the mouths of consumers. When we eat meat, the process of chewing initially involves the breakdown or fracture of the material due to the loads and deformations imposed on it by the teeth. The ease or difficulty with which these fracture processes are achieved is sensed by physiological transduction mechanisms and forms the sensory perception of tenderness or toughness as described by a taste panel. Unlike the precise designation of toughness (as resistance to the propagation of fracture) afforded by materials science definitions, toughness in the meat texture area is simply but imperfectly denoted as the description offered by sensory panellists. Other chapters in this volume cover the evolution of teeth to deal with various foods (Lucas), the mechanics of mastication (Heath), and to some extent the sensory aspects of the process (Lillford). In this chapter, attention will be concentrated on cooked meat as a material and how its structure dictates those of its fracture properties of relevance to this sensation of toughness. Because human consumption of meat is almost entirely in the cooked state, discussion of the material's mechanical behaviour will be restricted to meat after cooking. Toughness is perhaps the most important aspect of the eating quality of meat today (Lawrie, 1985) and possibly the least well understood and controlled. In the UK, meat is the most expensive food commodity in the household budget, accounting for about 30% of expenditure (Ministry of Agriculture, Fisheries and Food, 1989). Meat toughness is therefore an important issue in our current context and has been the subject of a large amount of research during this century.

The structure of meat

The general structure of meat is shown in Fig. 1. Surrounding each muscle is a connective tissue sheath, the epimysium, which is continuous with the tendon. Internally, the muscle is divided up into bundles of muscle fibres, each bundle being surrounded by the perimysium, a crossed-ply network of crimped, coarse collagen fibres. The mechanical properties of this network have been modelled (Purslow, 1989). Perimysia surrounding adjacent bundles merge to form a continuous network which is easily seen by eye when a transverse slice of meat is looked at. Fibre bundles are typically 1–5 mm across. Within the bundle, individual muscle fibres, typically 10–100 μm wide and up to a few centimetres long, are separated from each other by the endomysial connective tissue. Each muscle fibre is a multinucleated cell that contains about 1000 myofibrils, the contractile apparatus of the muscle. Myofibrils are roughly cylindrical, with diameters of 1–2 μm, and have a regular striated appearance due to the regularly repeating sarcomere units along their length. The ends of the sarcomere are bounded by the Z-disc. Emerging from the Z-disc are the thin filaments, composed principally of actin. Thin filaments interdigitate with the thick filaments found centrally in the sarcomere, which are composed principally of myosin. Myosin molecules have a long (156 nm) tail and two pear-shaped heads (19 nm long). The tails pack to form the shaft of the 1.6 μm long thick filament, leaving the heads on the surface of the filament. Thick filaments are laterally linked by the M-line half way along their length. Not shown in Fig. 1 are the gap filaments that contain the protein titin and run from the Z-disc to the M-line. Muscle contraction involves the cyclic interaction of myosin heads with the actin molecules in the thin filaments, triggered by calcium ions and fuelled by hydrolysis of ATP. After slaughter of an animal the ATP concentration

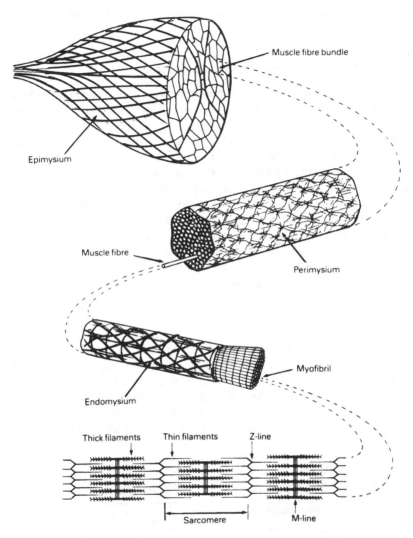

Fig. 1. The general structure of meat. (From Jolley & Purslow, 1988, by permission of the Publishers, Butterworth & Co. (Publishers) Ltd ©.)

eventually falls to zero in the muscle, at which time the myosin heads attach to the thin filaments, making the muscle stiff. This condition is known as *rigor mortis*. If the muscle is kept for several days after development of rigor, the stiffness gradually reduces due, it is thought, to the degradation of the myofibrillar structure by proteolysis. This process, known as ageing or post-mortem conditioning, has long been known to tenderise meat.

Toughness measurement devices

Since the turn of the century, a great many mechanical test methods have been proposed and used to measure some aspect of the breakdown or fracture of cooked meat that correlates with the sensory perception of texture. Szczesniak & Torgeson (1965) in their review of methods of meat texture measurement list approximately 20 basic types of 'shearing', penetration, biting, grinding and compression devices that have appeared in the literature, together with numerous sub-variants of each type. More devices and techniques have been proposed since 1965. The earliest reported device was that of Lehmann (1907): it was essentially a device to emulate the biting action of the incisors. The devices that have proved the most popular in the long run can be divided roughly into two main types, the so-called 'shear' tests and the bite tests. The essential features of these two test types are shown schematically in Fig. 2.

'Shear' tests

Possibly the most widely used shear device was first proposed by Warner (1928) and subsequently modified by Bratzler (1932), and is known as the Warner–Bratzler shear test. It consists of a thin blade with a triangular hole through it, into which a cylindrical sample of meat is placed. The blade is moved at a constant rate through a slot, so that the meat is caught across the edges of the slot and is ruptured or 'sheared' by the inside edge

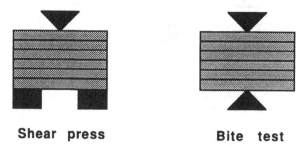

Shear press   **Bite test**

Fig. 2. Diagram showing the basic principles of shear and bite tests. In the shear tests a blunt wedge is driven through meat (shaded) supported on a slot through which the blade passes. Bite tests again have a blunt wedge or blade driven into the meat, but the meat is supported on a stationary wedge axially opposed to the driven one.

of the hole in the blade. The mode of deformation experienced by the sample is actually nothing like simple shear, but rather a complex and test-dependent mix of compression, shear and tension (Voisey, 1976; Harris & Shorthose, 1988). It is most difficult to visualise the deformation in the meat during this test, but Voisey (1976), using a transparent acrylate model of the apparatus, observed that the meat ultimately failed in tension with the muscle fibres stretched around the blunt edge of the blade. The popularity of the Warner–Bratzler technique lies in the degree of correlation that has been found between the maximum force recorded during the test with sensory panel scores of meat toughness. Szczesniak & Torgeson (1965) list 51 papers in which such tests have been carried out, of which 41 reported good or very good correlation. They partly attribute the poor and insignificant correlations in the remaining 10 papers to some doubts about the reliability of the sensory panels.

Variants on the shear test have been common. Kramer, Burkhardt & Rogers (1951) used a single flat-ended blade which ran through a slotted plate on which the meat sample sat. A variant of his, which used multiple (10 or 13) blades running through a slotted cage in which the sample is contained, is often referred to as the Kramer shear press (Voisey, 1976) or the L.E.E.–Kramer shear press (Szczesniak & Torgeson, 1965).

Bite tests

Most of the devices in this category can be viewed as derivatives of the Volodkevich bite tenderometer (Volodkevich, 1938). This consisted of two axially opposed flat blades with edges of constant radius, the bottom one remaining stationary and the top one being driven down towards it. A prismatic bar of cooked meat is placed between the two and the force required to 'bite' through it recorded. Figure 3 diagrammatically shows the Volodkevich-type jaws used in the Institute of Food Research Bristol Laboratory for meat toughness testing (Rhodes *et al.*, 1972). The MIRINZ tenderometer, derived from the device of Macfarlane & Marer (1966), used extensively in New Zealand to study factors affecting meat toughness, works on the same principle.

The load-deformation record from a bite test apparatus such as that shown in Fig. 3 mounted in an Instron mechanical testing machine is shown in Fig. 4. Two peaks in load are seen. It has been established that the height of the first peak (B, at the smaller deformations) varies most with treatments to the meat that preferentially affect the structure and integrity of the myofibrils. The second, final peak (peak A) varies most with treatments that primarily affect intramuscular connective tissue. Although reported Warner–Bratzler load-deformation curves are quite

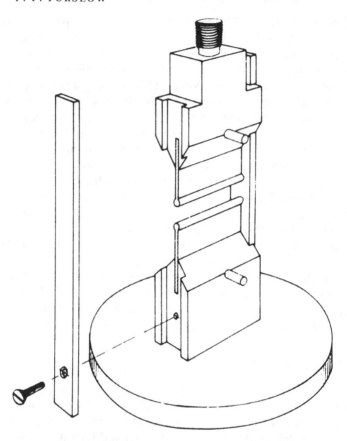

Fig. 3. Diagram of the Volodkevich bite test jaws used in conjunction with an Instron materials testing machine. (From Rhodes *et al.*, 1972, with permission, Kluwer Academic Publishers.)

variable (cf. Voisey, 1976; Møller, 1980–1; Harris & Shorthose, 1988), the same twin-component nature can be discerned. As with bite tests, the initial peak at smaller deformations is thought to correspond to the rupture of myofibrillar material, with the final peak being due to rupture of connective tissue. Møller (1980–1) stopped his Warner–Bratzler tests after the initial peak but before final rupture and observed connective tissue sheaths (most probably perimysium) holding the sample together. This common pattern of muscle fibre rupture at small extensions and subsequent connective tissue rupture in the shear and bite tests has important parallels in the tensile fracture behaviour of cooked meat, as will be described below.

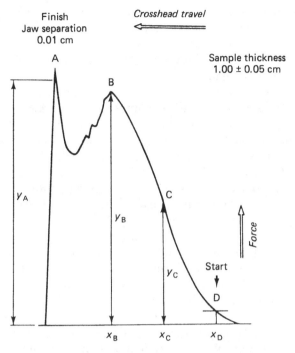

Fig. 4. Load-deflection diagram from test using apparatus in Fig. 3. (From Rhodes *et al.*, 1972, with permission, Kluwer Academic Publishers.)

We may digress slightly to consider another type of bite test which aimed to emulate the chewing action in the mouth more exactly. The MIT tenderometer (Proctor *et al.*, 1955) used vertical and horizontal reciprocating motions to simulate jaw movement during chewing cycles, based on previous dental studies, the forces being applied to the meat via a set of dentures. The more widely used General Foods texturometer (Friedman, Whitney & Szczesniak, 1963), described in detail elsewhere in this volume (Kilcast & Eves), was developed from the MIT device. It is interesting to note that the General Foods apparatus, preferred for its reproducibility and good correlations with sensory panel texture scores, is less imitative of the chewing process than the MIT device; it uses only a vertical reciprocating motion and a flat plunger is substituted for the dentures. Szczesniak (1987) reflected upon the apparent disparity in the modern approaches to measuring texture. On the one hand, the acceptance of texture as a complex, multivariate sensation leads to ever-increasing degrees of sensory profile analysis. On the other hand the recognition that texture can be adequately quantified by considering just the two essential parameters of toughness and juiciness leads to the simplification of the measurements required.

Whilst these bite and shear tests have been successful in predicting sensory panel scores of toughness, from a fundamental mechanical point of view these tests are unattractive. They are highly empirical, the actual values of peak force and energy used to rupture the meat depending on the precise construction of the apparatus; angle and radius of blade or wedge tip, thickness of blades, clearance between blades and slots in the shear press types all have large effects on measured values (see Voisey (1976) for a detailed critique). In the Volodkevich-type tests the moving wedge must be stopped before it makes contact with the stationary wedge to avoid damaging the apparatus; the height of the final load peak clearly depends strongly on the final jaw separation. Neither is it possible to measure any well-defined, basic mechanical properties from such tests. They can be used to map quickly and reproducibly the effects of treatments that affect meat toughness. However, they are of limited value in trying to understand how and why, on a detailed and fundamental cause-and-effect basis, variations in meat toughness come about.

To answer that type of question, we need to identify the sequence of structural mechanisms involved in the breakdown of cooked meat and to be able to quantify the overall contribution of each mechanism to overall fracture properties on the basis of well-defined mechanical parameters. By understanding how the relative balance between the various mechanisms is changed by factors such as cooking time and temperature or degree of post-mortem conditioning, we will move closer to an understanding of how changes in the structure of the material determine the fracture properties of meat, and therefore be in a better position to improve and control variations in meat texture.

Simple tensile tests have increasingly been used in this regard (e.g. Bouton & Harris, 1972a,b,c; Bouton, Harris & Shorthose, 1975; Davey & Gilbert, 1977; Locker & Carse, 1976; Penfield, Barker & Meyer, 1976; Stanley & Swatland, 1976; Carroll et al., 1978; Purslow, 1985). Tensile tests allow easier observation of what breaks where and at what stage when meat is pulled apart. They also give simple and unambiguous measurements of basic fracture properties such as breaking strength and breaking strain. In the remaining part of this chapter, the results from tensile tests will be used to demonstrate how far we have come in understanding the basic structural causes of the fracture behaviour of cooked meat.

Structural fracture mechanisms during tensile tests across the fibre direction

Let us examine the pattern of structural events that we can see when cooked beef muscle (*musculus semitendinosus*) is pulled transversely to the muscle fibre direction. Figure 5 shows the initial stages of such an experiment, at moderate extensions. Gaps open up in the bulk of the specimen at very precise locations – the junctions between muscle fibre bundles (Carroll *et al.*, 1978; Purslow, 1985). The perimysial connective tissue sheet (Fig. 5, double arrow) lying between the two bundles remains intact. The gap grows by rupture of the fine connectives (Fig. 5, single arrow) joining the perimysium to the endomysial connective tissue sheaths surrounding the muscle fibres on the surface of the fibre bundle. Scanning electron microscopy studies on the fractured surfaces from these kind of tests have directly confirmed this (Purslow, 1987; and, on a different muscle, Bernal & Stanley, 1987). Separation of the perimysial–endomysial junction is therefore the initial structural mechanism involved in the breakdown of meat pulled in this direction. Development of the final site of separation of the two parts of the specimen involves the

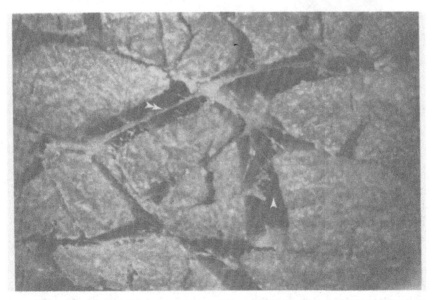

Fig. 5. Initial structural events in a transverse slice of cooked meat pulled across the muscle fibre direction. Gaps between muscle fibre bundles contain intact perimysial sheets (double arrow). Rupture of fine connectives (single arrow) results from gap opening.

Fig. 6. Later events in the breakdown of a transverse slice of cooked meat. Perimysial strands (arrowed) are the last structures to break.

joining up of many of these gaps between fibre bundles upon further extension of the test piece. This is shown in Fig. 6. The isolated but intact strands of perimysium (Fig. 6, arrowed) are left bridging the two sides of the macroscopic failure line and have to be broken for fracture to be complete.

These simple tests have thus revealed that two separate structural events involving intramuscular connective tissue are responsible for the fracture behaviour of cooked meat pulled across the fibre direction; perimysial–endomysial junction separation and rupture of isolated perimysial strands. Having qualitatively identified these two mechanisms, the next task is to quantify their relative contributions to the overall strength of meat pulled in this direction. This can be achieved by measur-

(a)

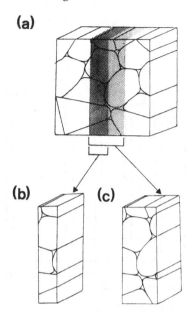

(b)   **(c)**

Fig. 7. Schematic diagram of cooked meat showing (a) the perimysial network across the transverse face of a meat block, and the sub-sectioning of this network in (b) thin and (c) thicker longitudinal slices. (From Lewis & Purslow, 1990, with permission.)

ing the effects of varying specimen size on the transverse strength of the meat (Lewis & Purslow, 1990), as explained schematically in Fig. 7. A block of cooked meat is viewed end-on to the muscle fibre direction in Fig. 7a. The black lines represent the perimysial connective tissue. If a thin longitudinal slice is taken along the muscle fibre direction, the sample with the cross-section shown in Fig. 7b is obtained. Pulling this specimen to breaking point perpendicular to the muscle fibre direction will merely necessitate separation of one or more of the perimysial–endomysial junctions for complete fracture to occur. If, however, a thicker longitudinal slice (Fig. 7c) is pulled perpendicular to the muscle fibre direction, sufficient perimysium is included in this sample to form an intact strand which runs the length of the perpendicular sample. In this thicker slice, then, this perimysial ribbon has to be ruptured after perimysial–endomysial separation has occurred for the specimen to be broken completely.

Examination of the variation in tensile strength of cooked meat pulled perpendicular to the fibre direction therefore allows the relative strengths of these two connective tissue fracture events to be quantified. Results

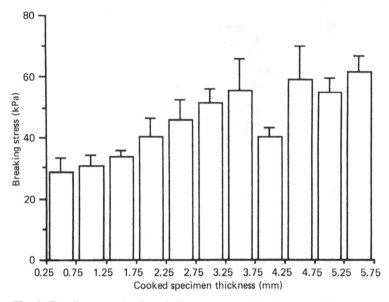

Fig. 8. Tensile strength of cooked meat perpendicular to the fibre direction as a function of longitudinal slice thickness. Bars denote one standard error. (From Lewis & Purslow, 1990, with permission.)

from such a series of tests are shown in Fig. 8 (Lewis & Purslow, 1990). In tests on the thinnest samples, where perimysial–endomysial separation is the only necessary event, tensile strength is seen to be quite low, about 30 kPa. As slice thickness increases, the probability that an intact perimysial ribbon is included in the specimen increases until, in the thickest slices, it is guaranteed. The tensile strength increases in line with this increasing probability, reaching an upper value of about 60 kPa. Perimysial strand rupture therefore requires higher stresses than perimysial–endomysial separation. Tensile tests on perimysial ribbons isolated from cooked meat (Lewis & Purslow, 1989) show that perimysial strength is maximum in meat cooked at 50 °C, decreasing at higher cooking temperatures.

The amount of intramuscular connective tissue and its degree of intermolecular cross-linking has long been known to be a strong factor in determining meat toughness (e.g. Strandine, Koonz & Ramsbottom, 1949; Dransfield, 1977; Bailey, 1985; Light, 1987). The investigations outlined above have begun to show why concentration is now being focused on the amounts and composition of specifically the perimysial connective tissue as the key structure in the connective tissue contribution to meat toughness (Light *et al.*, 1984).

Tensile tests along the muscle fibre direction

If we now consider what has to happen when cooked meat is pulled to fracture along the muscle fibre direction it is apparent that, as well as the perimysial connective tissue, all the fibres and fibre bundles in the cross-section necessarily have to be broken for complete fracture to occur.

Figure 9 shows schematically the load–extension behaviour of cooked meat pulled along the muscle fibre direction. The upper trace is for muscle that has been taken out of the carcase just after *rigor mortis* has been developed, and cooked. The lower trace is for muscle that has been post-mortem conditioned (aged) by being kept for a further two weeks post mortem before being cooked. The aged case is discussed later; the unaged trace will be evaluated first. For the unaged material, which is directly comparable to that used in the transverse orientation tests described above, the peak stress is about an order of magnitude higher than the maximum shown in Fig. 8 for tests across the fibre direction. The breaking strain is also high, often exceeding 100%. Cooked muscle fibres have higher ultimate extensions than raw fibres; although considerable variation is observed, in some cases they can stretch by more than 100% (Wang *et al.*, 1956; Hostetler & Cover, 1961). Isolated perimysium from cooked meat is also very extensible; again variability is considerable but in some cases extensions of 150% have been achieved (G.L. Lewis & P.P. Purslow, unpublished results). Meat cooked to 80°C is initially stiff

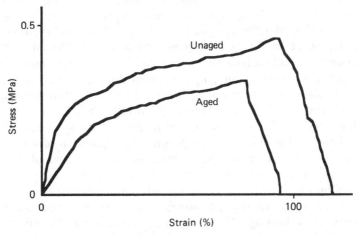

Fig. 9. Examples of the load–extension behaviour of cooked meat pulled to failure along the muscle fibre direction. Traces are shown for meat cooked 24 h post mortem (unaged) and 2 weeks post mortem (aged).

(up to extensions in the region of 10–20%); this is followed by a less stiff region, the transition often being referred to as the 'yield' or 'change' point (Bouton *et al.*, 1975; Locker, Wild & Daines, 1983). If the material is stretched beyond this region up to 65% extension and then the load is removed, the strip very nearly returns to its original length, but upon a second extension the first load-extension trace is not repeated; the load generated at low extensions is much smaller (Locker *et al.*, 1983). This indicates that irreversible breakdown of sub-structures occurs through and beyond the yield point. Locker *et al.* (1983) ascribed this to damage in the myofibrillar component. Cooked muscle still exhibits a striated appearance, but the thick filaments are thought to fuse into gel-like structures (Offer *et al.*, 1988). Locker and his collaborators have demonstrated that gap filaments may survive the cooking process (Locker *et al.*, 1977; Locker & Wild, 1982b). Prolonged cooking at 100 °C for 3 h, which is sufficient to solubilise much of the intramuscular connective tissue, diminishes the stiffness at high extensions and breaking strength but does not remove the yield point. This indicates the important role of connective tissue structures near the breaking point, but not in the yield point (Locker *et al.*, 1983).

If direct observations are made of cooked meat being extended, the precise sequence of events is more difficult to visualise than in transverse tests, but some general features can be discerned. Figure 10a shows a piece of unaged meat cooked to 80 °C with a single edge-notch in the left-hand side during the initial stages of extension along the fibre direction. The end of the notch is just beginning to take on a square appearance. This is consistent with the intact fibre bundles just ahead of the tip of the notch becoming laterally separated from the notched bundles, so allowing some degree of longitudinal slippage between them. At much greater extensions (Fig. 10b) it is obvious that extensive lateral separation of fibre bundles has occurred, opening up long splits along the fibre direction and perpendicular to the initial notch. These aspects of fracture behaviour of unaged meat have been observed previously (Carroll *et al.*, 1978; Purslow, 1985, 1987). The subsequent fracture of the specimen after the stage shown in Fig. 10b involved progressive separation of some of the bundles in the remaining web of the material. Separated bundles then fail independently. In such tests, isolated strands of highly stretched perimysium can sometimes be seen to stretch and break as the material nears its ultimate extension, although it is not clear if peak loads have been passed by this stage. Scanning electron microscopy investigations of the resulting fracture surfaces from such tests (Purslow, 1987) show well-separated bundles of muscle fibres, each isolated bundle having snapped across relatively cleanly. It is therefore apparent that fibre bundles and

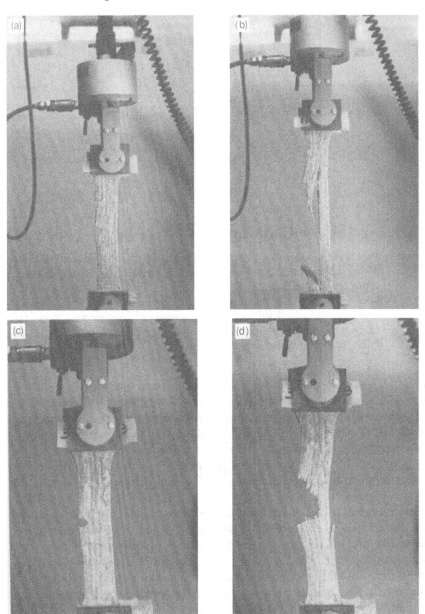

Fig. 10. The macroscopic appearance of the deformation and fracture of unaged and aged meat in tensile tests. Unaged and cooked meat at (a) early and (b) later stages of extension to fracture. Aged and cooked meat at (c) early and (d) later stages of extension.

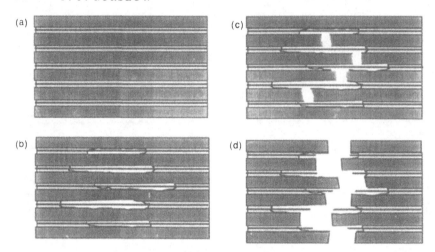

Fig. 11. Diagram of sequence of events during the extension to fracture of *unaged* cooked meat along the muscle fibre direction. Shaded bars represent the muscle fibre bundles, thick dark lines the perimysia between bundles and the cross-hatching represents the perimysial–endomysial junction. Four successive stages of extension, from (a) unstretched up to (d) total fracture, are depicted. (After Offer *et al.*, 1989, with permission.)

their surrounding perimysium are important structures in the fracture behaviour of meat pulled along, as well as across, the fibre direction.

Derived from the foregoing observations, a simple hypothesis for the sequence of events during extension along the fibre direction and fracture of unaged cooked meat has been proposed. This has been outlined briefly by Offer *et al.* (1988), but is here presented in more detail. The hypothesis is represented schematically by Fig. 11. In this diagram, shaded bars represent the muscle fibre bundles, thick dark lines the perimysia between bundles and the cross-hatching represents the perimysial–endomysial junction. Four successive stages of extension, from (a) unstretched up to (d) total fracture are depicted. It must be stressed that this is a very simplified diagram. At small extensions, the initial event shown in (b) is the separation of the perimysial–endomysial junction at the boundaries of the muscle fibre bundles, leaving isolated but intact muscle fibre bundles in parallel with isolated perimysia. At higher extensions individual fibre bundles snap somewhere along their debonded length, depending on the weakest point along this length (c), leaving the extensible, isolated perimysial strands as the last structures to break to effect complete fracture (d). Note that this sequence of muscle fibre fracture at lower extensions with subsequent connective tissue rupture concurs with the sequence thought to occur in shear and bite tests outlined above.

Tensile fracture of aged meat

Returning to Fig. 9, let us now examine the load–extension behaviour of aged and then cooked meat. The breaking strength along the muscle fibre direction declines with ageing, unlike the transverse breaking strength (Bouton & Harris, 1972c; Purslow, unpublished results). The initial stiffness for muscle aged for 7 days can decrease by as much as half (Locker *et al.*, 1983) and a decrease in the load and extension to the change or yield point is seen, although these decreases are by no means as dramatic as those seen in the yield point of raw meat on ageing (Locker & Wild, 1982*a*).

Ageing is known to degrade the myofibrils structurally. Extension of *raw* unaged muscle fibres occurs by a rather uniform but irreversible increase in sarcomere length (Offer *et al.*, 1988). However, raw aged muscle extends by myofibrillar fragmentation or gaps opening up within some sarcomeres. Within the sarcomere these micro-fractures occur mainly at the junction between the thin filaments and the Z-disc, but sometimes at the point where thin and thick filaments begin to overlap (Davey & Dickson, 1970; Locker & Wild, 1982*a*). We have proposed that if ageing uniformly weakens the thin filaments then these sites would be the most likely places for fracture to occur because they represent a rapid step-change in the amount of load-bearing material and so would be sites of stress concentration. This is especially true for the thin filament–Z-disc junction, the commonest site of fracture (Offer *et al.*, 1988).

Macroscopic observations of tensile tests on aged and then cooked muscle (Fig. 10c and d) are again not easy to interpret clearly, but some differences from the behaviour of unaged meat can be discerned. Figure 10c shows the initial stages of extension of a single edge-notched strip. At later stages (Fig. 10d) the progression of fracture across the specimen occurs at lower extensions than for the unaged material. Splits running along the fibre direction between muscle fibre bundles are not apparent, and the fracture runs more nearly in a straight line across the specimen, in line with the original notch. This is in stark contrast to unaged meat (Fig. 10b), where the resultant fractured surfaces appear very jagged, with isolated bundles protruding out and splits running back between them into the bulk of the specimen. This jagged appearance of fractured unaged specimens and more planar appearance of the fractured aged specimens is independent of the length of any single edge-notch cut in the specimen before loading; the length of protruding bundles and splitting

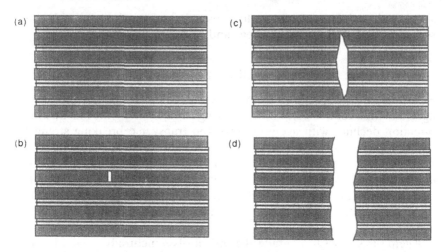

Fig. 12. Schematic diagram of sequence of events in *aged* muscle extended along the muscle fibre direction from (a) rest to (d) complete fracture. Fibre bundles, perimysia and perimysial–endomysial junctions are depicted as in Fig. 11.

back of fibre bundle junctions has been measured to be approximately three times longer in a group of unaged specimens than in a comparable group of aged specimens (Purslow, unpublished results).

In a manner similar to that of the scheme presented in Fig. 11 for the fracture of unaged meat, the foregoing observations on cooked meat can be used to hypothesise the likely course of events as aged meat is extended along the muscle fibre direction to fracture. This is shown schematically in Fig. 12, where four stages from (a) unstretched through to (d) complete fracture are depicted. Note that a sequence of events is proposed different from those in unaged meat. Because the muscle fibres are weakened but perimysial connective tissue and the perimysial–endomysial junction remain unaltered by the ageing process, the first event at small extensions may be the cracking or snapping of fibre bundles, shown in (b). The perimysial–endomysial junction is still intact and therefore the fibre bundles and the intervening perimysia are well-connected laterally. In this situation stress concentrations at the edge of the cracked bundle can exist, laterally transmitting the effects of the fracture of the first bundle to the intervening perimysium and the next fibre bundle, and so on across the material (Fig. 12c) causing the simultaneous failure of both components of the composite structure at stresses and strains below those necessary to break the unaged material, where no stress concentrations can exist because bundles and perimysia separate early on and subsequently fail independently. Propagation of fracture across the aged material due to stress concentration effects would create a more nearly planar fracture surface. The two hypotheses for the sequence of structural events during fracture of aged and unaged meat are presented as two extremes; the fracture behaviour of any meat sample is likely to fall somewhere in a spectrum of which Figs. 11 and 12 represent boundary cases, but tending towards the situation depicted in Fig. 12 with increased ageing.

The two types of composite behaviour proposed in Figs. 11 and 12 also have implications for the notch-sensitivity (see Jeronimidis, this volume) of the meat. However, it is clear that non-linear biological materials can deviate from the notch-sensitivity behaviour described by small-strain, linearly elastic theory (Purslow, 1991) and that the behaviour of cooked meat should not be assessed against the linear framework.

Clearly, a great deal of detailed information remains to be collected on the relative contributions of muscle fibres, fibre bundles and perimysium to the load–extension and fracture behaviour of muscle tissue aged for different times and cooked under a variety of conditions. However, the structural approach to meat texture described above is valuable in focusing attention on how the biochemistry, morphology and composition of intramuscular connective tissue and muscle fibre components ultimately determine the toughness of meat.

References

Bailey, A.J. (1985). The role of collagen in the development of muscle and its relationship to eating quality. *Journal of Animal Science*, **60**, 1580–7.

Bernal, V.M. & Stanley, D.W. (1987). Effect of cooking temperature on the fracture behaviour of pre-rigor bovine *sternomandibularis* muscle (Research note). *Canadian Institute of Food Science and Technology*, **20**, 56–9.

Bouton, P.E. & Harris, P.V. (1972a). The effects of cooking temperature and time on some mechanical properties of meat. *Journal of Food Science*, **37**, 140–4.

Bouton, P.E. & Harris, P.V. (1972b). A comparison of some objective methods used to assess meat tenderness. *Journal of Food Science*, **37**, 218–21.

Bouton, P.E. & Harris, P.V. (1972c). The effects of some post-slaughter treatments on the mechanical properties of bovine and ovine muscle. *Journal of Food Science*, **37**, 539–43.

Bouton, P.E., Harris, P.V. & Shorthose, W.R. (1975). Possible relationships between shear, tensile and adhesion properties of meat and meat structure. *Journal of Texture Studies*, **6**, 297–314.

Bratzler, L.J. (1932). *Measuring the tenderness of meat by means of a mechanical shear*. M.S. thesis, Kansas State College.

Carroll, R.J., Rorer, F.P., Jones, S.B. & Cavanaugh, J.R. (1978). Effect of tensile stress on the ultrastructure of bovine muscle. *Journal of Food Science*, **43**, 1181–7.

Davey, C.L. & Dickson, M.R. (1970). Studies in meat toughness. 8. Ultra-structural changes in meat during ageing. *Journal of Food Science*, **35**, 56–60.

Davey, C.L. & Gilbert, K.V. (1977). Tensile strength and tenderness of beef sternomandibularis muscle. *Meat Science*, **1**, 49–61.

Dransfield, E. (1977). Intramuscular composition and texture of beef muscles. *Journal of the Science of Food and Agriculture*, **28**, 833–42.

Friedman, H.H., Whitney, J.E. & Szczesniak, A.S. (1963). The texturometer – a new instrument for objective texture measurement. *Journal of Food Science*, **28**, 390–6.

Harris, P.V. & Shorthose, W.R. (1988). Meat texture. In *Developments in Meat Science – 4*, ed. R. Lawrie, pp. 245–96. Elsevier, London.

Hostetler, R.L. & Cover, S. (1961). Relationship of extensibility of muscle fibres to tenderness of beef. *Journal of Food Science*, **26**, 535–40.

Kramer, A., Burkhardt, G.J. & Rogers, H.P. (1951). The shear-press, an instrument for measuring quality of foods. I. The instrument. *Canner*, **112**, 34–6.

Lawrie, R.A. (1985). *Meat Science*, 4th edn. Pergamon Press, Oxford.

Lehmann, K.B. (1907). Studien über die Zähigkeit des Fleisches und ihre Ursachen. *Archiv für Hygiene*, **63**, 134–79.

Lewis, G.L. & Purslow, P.P. (1989). The strength and stiffness of perimysial connective tissue isolated from cooked beef muscle. *Meat Science*, **26**, 255–69.

Lewis, G.L. & Purslow, P.P. (1990). Connective tissue differences in the strength of cooked meat across the muscle fibre direction due to test specimen size. *Meat Science*, **28**, 183–94.

Light, N.D. (1987). The role of collagen in determining the texture of meat. In *Advances in Meat Research*, vol. 4 *Collagen as a Food*, eds. A.M. Pearson, T.R. Dutson & A.J. Bailey, pp. 87–107. Van Nostrand Reinhold, New York.

Light, N.D., Champion, A.E., Voyle, C. & Bailey, A.J. (1984). The rôle of epimysial, perimysial and endomysial collagen in determining texture in six bovine muscles. *Meat Science*, **13**, 137–49.

Locker, R.H. & Carse, W.A. (1976). Extensibility, strength and tenderness of beef cooked to various degrees. *Journal of the Science of Food and Agriculture*, **27**, 891–901.

Locker, R.H., Daines, G.J., Carse, W.A. & Leet, N.G. (1977). Meat tenderness and the gap filaments. *Meat Science*, **1**, 87–104.

Locker, R.H. & Wild, D.J.C. (1982a). Yield point in raw beef muscle.

The effects of ageing, rigor temperature and stretch. *Meat Science*, **7**, 93–107.

Locker, R. H. & Wild, D. J. C. (1982b). Myofibrils of cooked meat are a continuum of gap filaments. *Meat Science*, **7**, 189–96.

Locker, R. H., Wild, D. J. C. & Daines, G. J. (1983). Tensile properties of cooked beef in relation to rigor temperature and tenderness. *Meat Science*, **8**, 283–99.

Macfarlane, P. G. & Marer, J. M. (1966). An apparatus for determining the toughness of meat. *Food Technology*, **20**, 134–5.

Ministry of Agriculture, Fisheries and Food (1989). *Household Food Consumption and Expenditure 1987*. Annual Report of the National Food Survey Committee. HMSO, London.

Møller, A. J. (1980–1). Analysis of Warner–Bratzler shear pattern with regard to myofibrillar and connective tissue components of tenderness. *Meat Science*, **5**, 247–60.

Offer, G., Knight, P., Jeacocke, R., Almond, R., Cousins, T., Elsey, J., Parsons, N., Sharp, A., Starr, R. & Purslow, P. P. (1989). The structural basis of water holding, appearance and toughness of meat and meat products. *Food Microstructure*, **8**, 151–70.

Offer, G., Purslow, P. P., Almond, R., Cousins, T., Elsey, J., Lewis, G., Parsons, N., Sharp, A., Starr, R. & Knight, P. (1988). Myofibrils and meat quality. *Proceedings of the 34th International Congress of Meat Science and Technology (Brisbane)*, Part A, pp. 161–8.

Penfield, M. P., Barker, C. L. & Meyer, B. H. (1976). Tensile properties of beef *semitendinosus* muscle as affected by heating rate and end point temperature. *Journal of Texture Studies*, **7**, 77–85.

Proctor, B. E., Davison, S., Malecki, G. J. & Welch, M. (1955). A recording strain-gauge denture tenderometer for foods. I. Instrument evaluation and initial tests. *Food Technology*, **9**, 471–7.

Purslow, P. P. (1985). The physical basis of meat texture: observations on the fracture behaviour of cooked bovine *M. semitendinosus*. *Meat Science*, **12**, 39–60.

Purslow, P. P. (1987). The fracture properties and thermal analysis of collagenous tissues. In *Advances in Meat Research*, vol. 4 *Collagen as a Food*, eds. A. M. Pearson, T. R. Dutson & A. J. Bailey, pp. 187–208. Van Nostrand Reinhold, New York.

Purslow, P. P. (1989). Strain-induced reorientation of an intramuscular connective tissue network: implications for passive muscle elasticity. *Journal of Biomechanics*, **22**, 21–31.

Purslow, P. P. (1991). The notch-sensitivity of non-linear materials. *Journal of Materials Science* (in press).

Purslow, P. P. & Jolley, P. D. (1988). Reformed meat products – fundamental concepts and new developments. In *Food Structure – Its Creation and Evaluation*, ed. J. M. V. Blanchard and J. R. Mitchell, pp. 231–64. Butterworths, London.

Rhodes, D. N., Jones, R. C. D., Chrystall, B. B. & Harries, J. M. (1972). Meat texture. II. The relationship between subjective assessments and a compressive test on roast beef. *Journal of Texture Studies*, **3**, 298–309.

Stanley, D. W. & Swatland, H. J. (1976). The microstructure of muscle tissue – a basis for meat texture measurement. *Journal of Texture Studies*, **7**, 65–75.

Strandine, E. J., Koonz, C. H. & Ramsbottom, J. M. (1949). A study of variations in muscles of beef and chicken. *Journal of Animal Science*, **8**, 483–94.

Szczesniak, A. S. (1987). Sensory texture evaluation methodology. *Proceedings of the 39th Reciprocal Meat Conference (1986)*, pp. 86–95. US National Livestock and Meat Board, Chicago.

Szczesniak, A. S. & Torgeson, K. W. (1965). Methods of meat texture measurement viewed from the background of factors affecting tenderness. *Advances in Food Research*, **14**, 33–165.

Voisey, P. W. (1976). Engineering assessment and critique of instruments used for meat tenderness evaluation. *Journal of Texture Studies*, **7**, 11–48.

Volodkevich, N. N. (1938). Apparatus for measurement of chewing resistance or tenderness of foodstuffs. *Food Research*, **3**, 221–5.

Wang, H., Doty, D. M., Beard, F. J., Pierce, J. C. & Hankins, O. G. (1956). Extensibility of single beef muscle fibres. *Journal of Animal Science*, **15**, 97–108.

Warner, K. F. (1928). Progress report on the mechanical tenderness of meat. *Proceedings of the American Society for Animal Production*, **21**, 114.

F. A. SIBBING

Food processing by mastication in cyprinid fish

Aquatic food is diverse in size, shape, location, motility, chemical and mechanical properties (Nikolsky, 1963) as are adaptations in fish to capture and process food. Unlike mammals, fish have no claws, so food collection is performed entirely by the oropharyngeal apparatus. In some, mainly predatory, fish the mouth is specialised for fast and voluminous suction. In others, mainly planktivorous fish, prey capture proceeds by overswimming, and suction plays a minor role (van Leeuwen & Muller, 1984; Sibbing, 1991). The complex of muscles, bones and ligaments composing the suction-pressure pump of the respiratory apparatus (cf. Ballintijn, 1969) eventually performs a whole range of diversified feeding actions. By accurate timing of volume changes in the oral, buccal, pharyngeal and opercular cavities and by differential operation of the oral and opercular valves, carp can perform as many as ten stereotyped patterns of movement, each serving a particular role in food processing (Sibbing, Osse & Terlouw, 1986; and see Fig. 1). These patterns are: (1) particulate intake or (2) gulping for intake, (3) rinsing, (4) spitting or (5) selective retention for sorting food from non-food, (6) transport, (7) loading, (8) crushing, (9) grinding and (10) deglutition or swallowing. In addition, size-selective retention of small food particles (sieving) is effected by the branchial sieve (Zander, 1906; Hoogenboezem et al., 1990), if such particles are abundant. Whereas size, location and motility of food are crucial factors for capture strategies, taste, degree of mixture with non-food and size are determinants in sorting out food from non-food (Sibbing et al., 1986). But how do fish cope with the highly varied mechanical properties of food?

The ability of animals to utilise ingested food resources depends on their ability to comminute them, increasing the surface area for digestive enzymes, or merely breaking indigestible capsules. Since plant materials impose extraordinary demands on mechanical (Vincent, this volume) and chemical breakdown by fish (cf. Prejs, 1984; Hofer, 1991) it is not surprising that, even in tropical zones with continuous crops, very few fish are

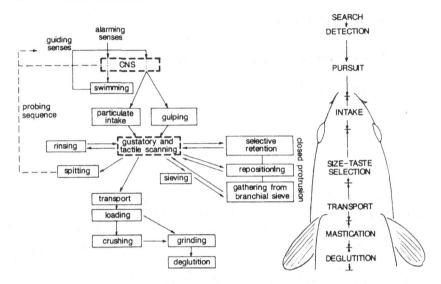

Fig. 1. Interrelations between distinct components of food handling in carp. Their actual site of action in the fish is indicated at the right. These stereotyped patterns are variously integrated into effective feeding sequences, adjusted to the type of food (Sibbing *et al.*, 1986). 'Probing' refers to repetitive intake and spitting of bottom debris while evaluating its profitability. CNS, central nervous system.

true herbivores. The oropharyngeal cavity of primitive teleost fish was presumably covered with numerous small tooth plates used for seizing and transport of large fleshy prey (Nelson, 1969). In these predaceous fish, fusion subsequently led to enlarged tooth plates (e.g. in *Elops*) with a grasping type of dentition in areas of particular functional significance such as at the oral jaws (for seizing prey), in the basibranchial and palatal area (for prey transport) and eventually at the posterior gill arches (for swallowing) (Nelson, 1969). In these fish there was not much size reduction or chewing of the food. In higher teleosts such tooth plates fused to the gill arches which posteriorly even developed into pharyngeal jaws. Dentition became specialised for mastication in such pharyngognath fish (Liem & Greenwood, 1981; Lauder, 1983a). Thus previously unaccessible food types (e.g. snails, seeds) became utilisable. Some Characidae and Mugilidae even developed thick-walled gizzard-like stomachs for triturating algae with weak cell walls (Castro, Sasso & Katchburian, 1961; Daget, 1962). The eventual solutions in fish for exposing diversified materials to digestive enzymes are bound to the historical constraints of the group. In some groups (e.g. characids such as piranhas) oral biting predominates (Fig. 2a), in others (e.g. cyprinids) oral teeth are lacking and pharyngeal mastication is highly developed (Fig. 2b). In more advanced teleosts both oral biting and pharyngeal mastication occur (Liem & Greenwood, 1981) e.g. in cichlids (Freyer & Iles, 1972; Liem, 1973; Hoogerhoud & Barel, 1978; and see Fig. 2c). Although cichlid fish show a wide range of adaptations in both oral and pharyngeal jaw apparatus (Barel *et al.*, 1977), it is expected that cyprinid fish, especially, show most distinctly relations between food texture and their pharyngeal jaw system. This largest freshwater fish family (2000 species; Nelson, 1982) characteristically lacks oral teeth as well as a stomach with its aggressive chemical breakdown. Additionally, cyprinid fish lack cellulases (Hofer, 1991). As a result high demands are imposed on the pharyngeal jaw system in this family.

The following questions will therefore be addressed in the subsequent sections and exemplified by the pharyngeal jaw system of cyprinids.

1. What functional demands do aquatic food types impose on mastication in fish?
2. How does the masticatory apparatus in fish work?
3. How are its structure and function related to food texture?
4. Is the masticatory process adjustable to the type of food?
5. Does mastication dictate the food niche?
6. Are there true carnivores, herbivores and omnivores among fish?
7. Does mastication widen the exploitation of aquatic food resources?

Functional demands on mastication in fish

Animals must detect and collect their food and break it down into digestible components. Generally, mastication should increase the digestible area of the food by size reduction or by merely breaking open indigestible capsules. In order to be efficient a certain increase of digestible area should involve minimal costs of energy and time. This holds for the whole sequence of masticatory actions, which includes (1) food supply, freed from water, (2) transporting food to the working surfaces of the teeth (loading), (3) fracturing the food, (4) retaining particles between the teeth for repetitive chewing and (5) clearing the system of particles small enough to be digested. The need for an accurate adjustment of these actions in view of an efficient total sequence will be evident and discussed later (see Mastication as a composite process, below).

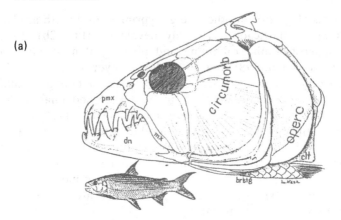

(a)

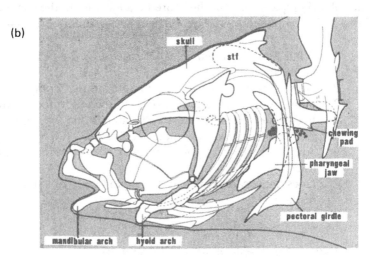

(b)

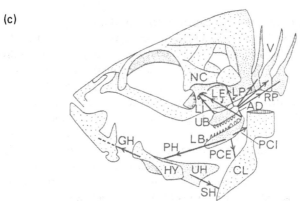

(c)

Food properties

The mechanism of food fracture and its efficiency depend largely on the structure and movement of the teeth and on the mechanical properties and texture of the food. Studies on mammalian mastication (Rensberger, 1973; Lucas, 1979, 1982; Lucas & Luke, 1984; Hiimae & Crompton, 1985) showed that knowledge of mechanical properties of biological materials as well as insight into how they break are bottlenecks in explaining relations between tooth design and diet. In addition, such relations in living organisms will be amended by historical and constructional constraints (cf. Barel, 1983) and by demands imposed by the other masticatory actions (see above).

Four classes of food texture are generally accepted (Lucas & Luke, 1984; Hiimae & Crompton, 1985): (1) hard brittle, (2) turgid, (3) soft tough and (4) tough fibrous (Table 1). Soft brittle materials (e.g. jellyfish) are weak and rare among biological tissues. Since we need unambiguous terms, their definition (Gordon, 1976; Alexander, 1983) is briefly summarised. *Hardness* (probably really stiffness) is a measure of deformability (elastic or plastic) and expressed as stress per unit of strain (Young's modulus, E). Hard (stiff) materials oppose *soft* (pliant) ones. *Brittle* items do not yield (plastic irreversible deformation) and are easy to crack. They contrast with *ductile* materials, which do flow beyond their elastic limit. Such ductile materials in this way resist further propagation

---

Fig. 2.(a) Characid fishes like this piranha (*Hydrocyon lineatus*) have strongly developed oral jaws and teeth (after Gregory, 1933). brstg, branchiostegal ray; clt, pectoral girdle; dn, dentary; mx, maxillary; pmx, premaxillary.

  (b) Head skeleton of common carp (only left side shown, gill covers and circumorbitalia removed). The fifth branchial arches have been modified into firm pharyngeal jaws bearing pharyngeal teeth medially. They oppose a horny chewing pad, fixed in the base of the skull. Cyprinids lack any other, including oral teeth. (Modified from Ballintijn, 1969) Stf: subtemporal fossa in the skull.

  (c) In cichlid fish the pharyngeal jaw system is composed of lower (LB) as well as upper (UB) pharyngeal jaws. The latter articulate with the skull. Oral teeth are present. Abbreviations refer to head muscles and bony elements (after Liem, 1973): AD, adductor; CL, pectoral girdle; GH, geniohyoideus; HY, hyoid; LE, fourth levator externus; LI, levator internus; LP, levator posterior; NC, neurocranium; PCE, pharyngocleithralis externus; PCI, pharyngocleithralis internus; PH, pharyngohyoideus; RP, retractor pharyngeus superior; SH, sternohyoideus; UH, urohyal; V, vertebrae.

Table 1. *Main categories of natural food, their mechanical properties and appropriate communition machinery. Arrows indicate the transition from macro- to microdiminution (modified from Lucas & Luke, 1984)*

	Hard brittle	Turgid	Soft tough	Tough fibrous
Examples	Seeds, nuts Unripe fruit Bone Some root storage organs Some adult insects	Ripe juicy fruits Some insect larvae	Animal soft tissues Some insects Leaves? Grasses?	Grass, fruit skin
Deformability	Stiff elastic	Plastic flow	Viscoelastic pliant	Viscoelastic pliant
Strength	High	Low	Moderate	Variable, according to fibre direction
Toughness	Low	Variable	High	High
Notch-sensitivity	High	Variable	Low	Low
Appropriate masticatory operation	Crushing, splitting →grinding	Crushing →grinding	Cutting, shearing, piercing	Cutting, lacerating, shearing
Machinery	Mortar–pestle	Mortar–pestle	Blades	Serial array of low profile blades

of cracks. This makes them *tough* – a measure of the work needed for fracture. *Strength* indicates the largest stress (tension, compression or shear) before breaking and is difficult to define for biological materials. Strong opposes *weak*.

Biological materials are generally composite and heterogeneous, which makes direction and velocity of crack propagation difficult to predict. Apart from the component material properties their structure is an important determinant for fracture. Fibrous materials can be notch-insensitive (Jeronimidis, this volume), cracks are readily absorbed at the matrix-fibre interface. Non-fibrous brittle materials are sensitive to small surface irregularities. Such notches produce stress concentrations which propagate cracks rapidly.

Masticatory operations

How do animals cope with such heterogeneous material as natural food? The animal has a series of masticatory operations as the main options to deform the food, starting simply from two approaching flat surfaces in crushing (Fig. 3). The size of the food–tooth contact area determines, at a given force and food, the magnitude of stresses in the food. For hard food this contact area depends mainly on its shape and position. Soft materials are compliant and crushing would readily spread the force over an increased contact area producing lower stresses for initiating cracks in the food. In addition crushing does not allow the tooth to penetrate the material in order to propagate cracks internally. The latter is required for fracturing tough and fibrous (notch-insensitive) items. For fracturing compliant or fibrous materials local deformations and stresses are increased by reducing the dimensions of the contact area from crushing towards splitting towards piercing (Fig. 3). A serial array of such instruments would compensate for the loss in working area and their spacing should be adjusted to the required particle size. Introducing an additional movement component parallel to the occlusal surfaces increases the

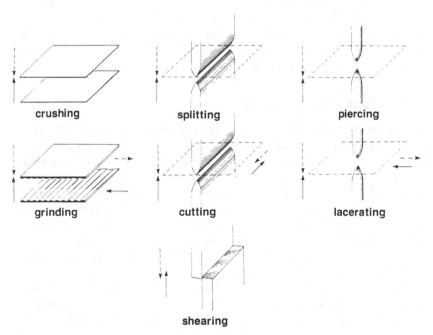

Fig. 3. Masticatory operations defined by the dimensions and movement (arrows) of two opposing surfaces (see the text).

contribution of shear in the external loading of the food. The contact area decreases from grinding towards cutting towards lacerating (Fig. 3).

The capacity for deformation is limited by the size, shape and range of movement of the opposing teeth. Local deformation and stress are greatly increased by adopting a set of opposing blades and aligning them so as to pass close to each other (shearing). In practice masticatory systems in animals combine several operations. In mammals three characteristic designs are as follows (Lucas & Luke, 1984; Hiimae & Crompton, 1985). (1) The pestle–mortar system for effectively crushing and grinding hard brittle (e.g. nuts) or turgid foods (e.g. juicy fruits) over a large area. Large particle fragments will be repetitively hit in the same chewing stroke. The marginal ridges surrounding such teeth may also aid in splitting and cutting. Pestle–mortar systems as such have not been found among fish. (2) Sharp blades (e.g. carnassials in carnivores), arranged in opposing pairs allowing splitting, cutting and shearing of soft and tough materials (cf. Table 1). Such systems are found in oral jaws of carnivorous piranhas as well (Goulding, 1985). (3) Serial arrays of low profile blades can fracture tough fibrous materials by combining shearing, cutting and lacerating in a type of milling machine, as seen in ungulates. No reports of such systems in fish have been found. It should, however, be noted that, whereas mammalian jaws combine oral biting and mastication, these are separate functions and structures in fish. A constructional demand of all such systems is that the reaction forces on teeth and jaws are readily absorbed and will not cause their fracture!

Macro- and microdiminution

Food items often have capsules (e.g. nuts, seeds) or skins (e.g. apple, fig, insects) which differ greatly in material properties and structure from their contents. This imposes an additional set of demands on the comminution machinery, just as the difference between very large and small particles does. *Macrodiminution* therefore refers to the initial stage of food breakdown at the organism–organ level preparatory to *microdiminution*, at the tissue–cell level. Heterodont dentition with a different design between anterior jaw teeth and posterior molars (e.g. in dogs), should be viewed in this context. Similar features are observed in fish and discussed later (p. 82). In some cases (e.g. snails) macrodiminution may require the most work and dominate the constructional design, in other cases (e.g. breaking down cell walls of small algae) microdiminution predominates. Breaking down large leaves imposes demands at both levels. The basic principles of mastication are based on investigation of mammals but how do fish break their food and what does their extremely varied tooth design (cf. Heckel, 1843; Chu, 1935; Vasnecov, 1939; Barel *et al.*, 1977) mean?

Pharyngeal mastication in fish

Pharyngeal jaw systems

Mastication demands larger forces than those required for transporting food. The evolutionary development of pharyngeal mastication in fish (see p. 58) is accompanied by a reduction in number but increase in size of the posterior branchial arch elements and their associated musculature. The levator muscle of the fifth branchial arch (LAB V) in carp for example (Fig. 4) is one of the heaviest head muscles and extends deep into the subtemporal fossa of the skull. Cyprinids developed a retractor muscle of the pharyngeal jaw (ROPI), which is derived from the oesophageal wall and is unique among lower teleosts. This muscle (Fig. 4) appears to play a crucial role in carp by adducting the pharyngeal jaws and by guiding crushing into grinding. The complete reduction of posterior upper branchial elements and the development instead of a bony

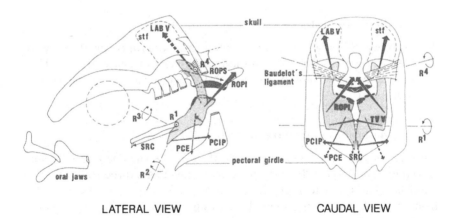

LATERAL VIEW   CAUDAL VIEW

Fig. 4.   Pharyngeal mastication apparatus of common carp (cf. Fig. 1b). Branchial arches and pectoral girdle have been partly cut away. Arrows indicate muscles, their thickness reflects the muscular weight. LAB V, musculus levator arcus branchialis V; PCE, m. pharyngo-cleithralis externus; PCIP, m. pharyngo-cleithralis internus posterior; ROPI, m. retractor os pharyngeus inferior; ROPS, m. retractor os pharyngeus superior; SRC, m. subcualis rectus communis; stf, subtemporal fossa in skull; TV V, m. transversus *v*; R^1–R^4, axes of rotation of the pharyngeal jaws.

anvil at the base of the cyprinid skull is a further step towards increasing loads within the pharyngeal jaw system. The action of teeth directly against the skull even resembles the mammalian design. The versatility of pharyngeal jaw movements is maintained by their movable interconnection and their free suspension in muscular slings to the skull and pectoral girdle. The only 'articulation' (R^1 in Fig. 4) is their connection to the fused basibranchials 3–4 in the floor of the branchial basket.

In higher teleosts similar trends evolved along a very different evolutionary line. In cichlid fish, upper branchial elements hypertrophy as well, forming the upper pharyngeal jaws (Fig. 2c; Liem, 1973). Each has an articular surface with the skull. The lower pharyngeal jaws fuse and function as a single unit against the upper pharyngeal jaws. The cichlid type of pharyngeal jaw system probably combines less high loadings with even higher versatility in movement, compared to cyprinids. Retractor muscles of the pharyngeal jaws developed independently from the branchial musculature. The posterior position of the pharyngeal jaws, close to the oesophagus, probably interferes less with suction and respiratory movements of the branchial basket. In addition mastication can occur after the water has been ejected through the branchial slits.

Pharyngeal mastication in common carp

A case study (Sibbing, 1982) on common carp, *Cyprinus carpio*, showed the basic design of the comminution machinery in cyprinid fish. This serves as a model for comparisons with trophic specialists in this family (see pp. 82–6).

Masticatory operations in carp

Occlusal profiles. The pharyngeal teeth and chewing pad of carp are shown in Figs. 5 and 12b. Each pharyngeal jaw bears three rows of teeth medially. The tooth formula of carp is 1 : 1 : 3 to 3 : 1 : 1. Whereas the front teeth (A1) are dome-shaped with a slight cusp, all others are molariform, i.e. flattened with sinusoidal furrows which will improve grip. All teeth oppose a horny chewing pad, showing the wearing facets and sliding movements of the respective teeth (Fig. 5; Sibbing, 1982). No traces of left–right tooth contact are apparent in carp. This combination of occlusal profiles allows for crushing and grinding, depending on the actual movements of the teeth during mastication. Splitting, cutting, lacerating and shearing are excluded in this design.

Occlusal movements. Measurement of the actual occlusal movements

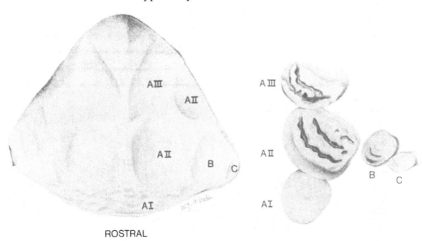

ROSTRAL

Fig. 5. Ventral view of chewing pad in carp. Left pharyngeal teeth for comparison at the same scale. Note the pitted anterior surface of the chewing pad reflecting the crushing action of the AI teeth and the grinding facets of the molariform teeth.

during feeding should elucidate their direction and amplitude and thus establish which masticatory operations occur. X-ray ciné film of carp pharyngeal jaws (Fig. 6) marked with platinum bars allowed such detailed measurements (Sibbing, 1982). The visualisation of these events (Fig. 7) has been reconstructed from movement graphs of the separate elements (Fig. 8). They show each masticatory cycle to be composed of a preparatory, a power and a recovery stroke. Two types of masticatory cycle are distinct. The powerstroke of the crushing cycle is characterised by adduction of the pharyngeal teeth, at the same time moving right on to the chewing pad (crushing). In the grinding cycle the teeth move apart (abduction) parallel to the chewing pad, perpendicular to the furrows in the molariform teeth. Both movement analyses (Fig. 8) and simultaneous electromyograms of the masticatory muscles (Figs. 9 and 11) show that crushing cycles always precede grinding cycles during the masticatory sequence. Intermediate cycles with evident crushing, as well as grinding phases, occur in the powerstroke. These data show that crushing and grinding are integrated into one system, the front teeth being differently shaped and operational in a different direction compared to the posterior teeth. In the time course of the masticatory sequence, frequency as well as amplitude of the masticatory cycles decrease (Fig. 9), suggesting that less power is needed for mastication of smaller bits.

(a)

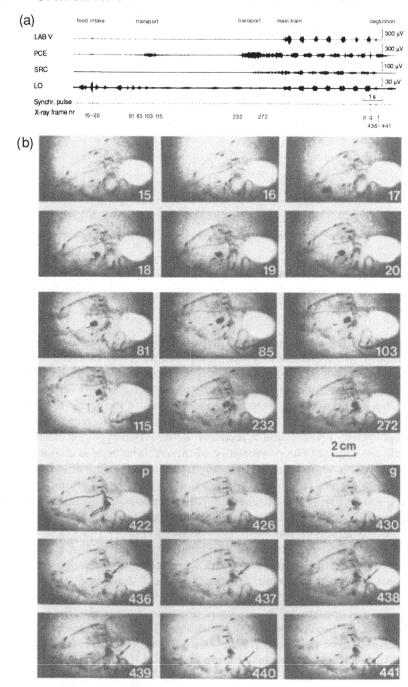

(b)

2 cm

Predicted and actual diet

Crushing serves macrodiminution of hard brittle (cf. Table 1) aquatic food items (e.g. seeds, bivalve and gastropod shells) and it will squeeze crustacean zooplankters and insect larvae ('fluid filled bags' according to Lucas (1982)). The slightly elevated cusp on the crushing teeth (A1) provides a small contact area for breaking even strong particles. (The crushing of dried maize can even be heard outside the experimental tank!) The high reaction forces are spread over the large tooth cross-sectional area, preventing tooth fracture. Reaction forces will be further spread over the pharyngeal jaws and over the basioccipital process of the skull by bony ridges. These ridges roughly reflect the direction of highest loadings.

Grinding serves microdiminution of such hard, brittle and turgid materials over a larger area. Stresses high enough to rupture cellulose cell walls are not expected to occur. Soft and tough materials will be mainly flattened, just like fibrous items. The actual diet of carp (crustacean zooplankters, chironomid larvae and the epifauna between littoral vegetation such as molluscs, copepods, trichopteran larvae) matches its masticatory abilities (Sibbing, 1988). Mismatches such as herbivory in carp are reported only when animal prey is scarce. Herbivory hardly allows growth of carp (Fischer, 1968). A positive selection of thin-shelled over thick-shelled molluscs of the same genus (Stein, Kitchell & Knezević, 1975) seems to exemplify the optimal foraging theory (Werner & Hall, 1974). Since shell thickness is only one of the many food parameters and mastication only one of the many feeding actions, masticatory abilities do not necessarily predict the food niche of fish (cf.

Fig. 6.(a) Electromyographic record of food processing in carp depicted in (b) (numbers refer to the X-ray frames). (b) X-ray cinerecording of the successive feeding actions (26 frames s^{-1}). Black bars are measuring markers. Feeding starts with rapid intake (60 cm s^{-1}) of the radio-opaque ($BaSO_4$-impregnated) food pellet by suction. Note the expanding buccopharynx (frames 15–20). Later, the food is propelled by peristalsis through the anterior pharynx to the rostral margin of the chewing pad (frames 81–115), and finally loaded into the expanding posterior pharynx, between teeth and chewing pad (frames 232–72), prior to the preparatory stroke of the first masticatory cycle. After the last grinding cycle (frames 422–30) the masticatory train is completed by deglutition (frames 436–41; see arrows). The remainder of the pellet stays at the rostral margin of the chewing pad. For abbreviations, see Fig. 4; LOP, levator operculi; p, preparatory stroke; g, grinding stroke. (From Sibbing, 1982.)

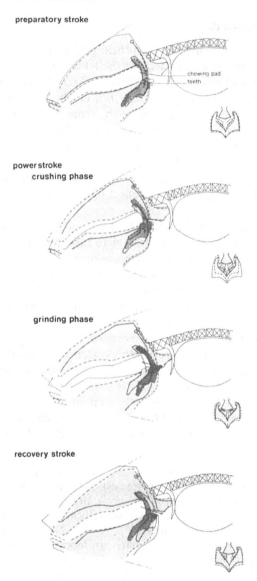

preparatory stroke

chewing pad
teeth

power stroke
crushing phase

grinding phase

recovery stroke

Fig. 7.   A grinding cycle of carp in lateral view, based on X-ray images. The insets offer a half-sized rostral view on the pharyngeal jaws and chewing pad to show adduction or abduction. For each of the four strokes, start (solid lines) and final position (dashed lines; dark field) are indicated. Final positions correspond to starting positions of the next stroke. The lines marking the pharyngeal lumen indicate bony parts; the actual lumen is, at the end of the preparatory stroke, fully occupied by

Mastication and ecological types, p. 86). For example, some snails may be easily crushed but the high shell : flesh ratio makes them less profitable (Hoogerhoud, 1987). Hoogerhoud demonstrated that ingestion of a high proportion of shell fragments in cichlids confers extra costs for their negative buoyancy and it consumes much of the intestinal storage capacity. This eventually increases the energetic costs and decreases the rate of feeding.

Machinery driving the occlusal surfaces

How are such different movement patterns as are required for crushing and grinding achieved in the pharynx of fish? Given the evolutionary connection of the pharyngeal jaws to the branchial basket they should roughly follow its movements. Only little forward and backward translation within this basket is allowed. Moving the teeth upward to the chewing pad for crushing is achieved by anticlockwise rotation of the pharyngeal jaws around their only joint-like connection, between their symphysis and the branchial basket (R^1 in Fig. 4a). Actually, muscles operating the pharyngeal jaws in crushing (Fig. 4, LAB V, ROPI, PCIP) have working lines that form effective couples around R^1. Teeth movements parallel to the chewing pad, required for grinding, are induced by Baudelot's ligament, which runs between the skull and the pectoral girdle (Fig. 4). Upon increasing retraction of the pharyngeal jaws by the ROPI their dorsal processes readily abut against these ligaments, which then become the centres of rotation (R^4). The ROPI muscles become effective rotators. Owing to the position of the pharyngeal teeth, close to R^4 during grinding, they make small but powerful caudad excursions across the chewing pad. In addition the pharyngeal jaws perform intrinsic movements around their movable symphysis. Rotation around this axis R^3 effects abduction of the jaws and teeth. Around the length-axis (R^2) of each jaw further adjustment of tooth position towards and away from the chewing pad can occur. Excursions of the teeth around R^1 and R^3 are large, due to their large distance from the centre of rotation. This allows much space (about 12 mm in a 300 mm carp) between the teeth and the

---

palatal and postlingual organs. Note the lifting of the pharyngeal floor and jaws in the crushing phase, whereas the chewing pad remains in about the same position. In the grinding phase the teeth and chewing pad move in opposite directions, wedging the food. Rotations of pharyngeal jaws are distinct; pure retraction is hardly noticed. Skull and pectoral girdle movements contribute conspicuously to grinding; the pharyngeal cavity is widely expanded. The picture of a crushing cycle differs by absence of a distinct grinding phase. (From Sibbing, 1982.)

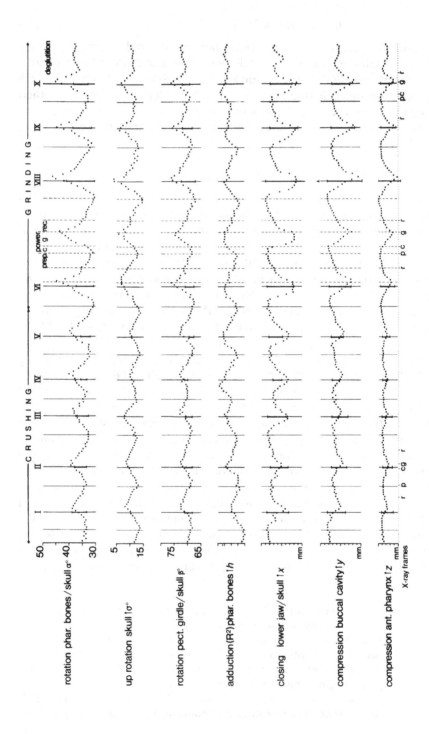

rotation phar. bones / skull $\alpha°$

up rotation skull $\uparrow\sigma°$

rotation pect. girdle / skull $\rho°$

adduction (R^2) phar. bones $\uparrow h$

closing lower jaw / skull $\uparrow x$

compression buccal cavity $\uparrow y$

compression ant. pharynx $\downarrow z$

X-ray frames

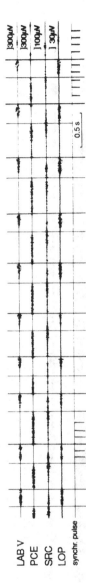

Fig. 8. Cyclic movements of head elements during the first train of a masticatory sequence in a feeding carp (pellets), measured in lateral X-ray films. Simultaneous electromyographs (EMGs) of selected pharyngeal jaw muscles permit comparison with the overall EMG pattern. Vertical lines mark termination of LAB V or PCE activity. First series (I–V; small amplitudes) represent crushing cycles, second series (VI–X; large amplitudes) grinding cycles. Cycle VII is subdivided into a preparatory (prep), crushing (c), grinding (g), and recovery (rec) stroke. In grinding the crushing phase is shortened and the grinding phase extended, compared to crushing (cf. cycle II). Note the large share of skull rotation in most movements, indicated by heavy bars on the vertical lines. Expansion of the buccopharyngeal cavity is extensive and synchronous with the power stroke. For abbreviations, see Figs. 4 and 6. (From Sibbing, 1982.)

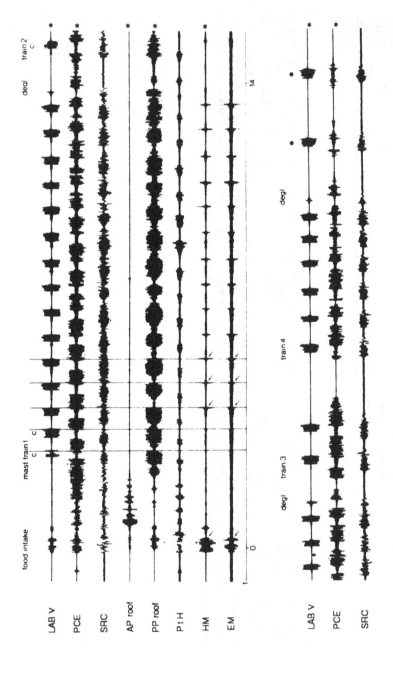

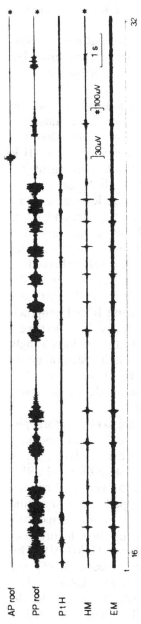

Fig. 9. Joint action of pharyngeal jaw, head, and body muscles during transport and mastication of a pellet in carp. The masticatory sequence is composed of four trains and some isolated cycles. First train includes a series of crushing (c) and a series of grinding cycles, and is completed by deglutition. The power stroke corresponds to the period of LAB V activity (upper trace). Note the transition of activity from the anterior to the posterior pharyngeal roof (palatal organ) after food intake, indicating its role in transport. Body muscles participate in food intake and grinding (arrows), whereas at crushing their activity is low. For abbreviations, see Fig. 4; AP, anterior pharyngeal; PP, posterior pharyngeal; PtH, protractor hyoideus; HM, hypaxial muscles; EM, epaxial muscles. (From Sibbing, 1982.)

chewing pad for food supply and increases the occlusion velocity of the teeth in crushing. High occlusion speeds would reduce the energy needed to fracture viscoelastic materials by not allowing time for viscous deformation in the food (Alexander, 1983).

Excursions around R^4 and R^2 are smaller and slower but more powerful, due to their small distance from the rotational axis, which is crucial in grinding. Tooth movement is perpendicular to the furrows in their crowns, thus improving grip for transport.

Resolving the action lines of each muscle into components around the separate axes, R^1, R^2, R^3 and R^4, predicts the main role of each muscle in these occlusal movements (Sibbing, 1982). The mass of the muscles indicates the available work per contraction, which is largest for the LAB V and ROPI muscles (respectively 33% and 38% of all pharyngeal jaw muscles together). The rotational moment of this LAB V for crushing (around R^1) by its large cross-sectional area and its huge lever exceeds by far (at least about 18 times) all other muscles. Grinding (around R^4) is most effectively performed by the ROPI and PCIP muscles (see legend to Fig. 4 for abbreviations). Dorsad positioning of the occlusal surfaces (around R^2) and adduction (around R^3) is most effectively achieved by the ROPI, ROPS and TV V muscles. Long thin muscles bring the unloaded pharyngeal jaws back into their rest position (PCE, SRC and ROPS) during the recovery stroke. Abduction around R^3 will lengthen the grinding stroke and results mainly from PCE activity. It is, however, also induced by the ligaments of Baudelot, which run increasingly caudolateral due to retraction of their attachment with the pectoral girdle during grinding. Thus PCIP action forces the pharyngeal jaws into abduction. Electromyograms confirm the activity of these muscles in the power stroke of mastication (Figs. 9 and 11; Sibbing, 1982).

Even though pharyngeal jaw muscles running from skull, pectoral girdle and branchial basket form effective rotation couples, thus together increasing the masticatory force, body muscles appear to play a crucial role in powering this system. Epaxial body muscles rotate the skull with a substantial lever upward around its connection to the vertebral column (R^v in Fig. 10). They thereby move the chewing pad in the direction opposite to the teeth rostro-ventrad, thus increasing shearing forces and wedging the food during grinding. Lifting the skull also lifts the origin of the powerful levator muscle of the pharyngeal jaw (LAB V). This increases the force in mastication by the following measures. (1) The force of the epaxial muscles is positioned in series with LAB V. This muscle has highly developed tendinous parts. (2) According to the length–tension diagram, production of forces will decrease if LAB V shortens. As soon as occlusion is maximal, prolonged skull rotation will

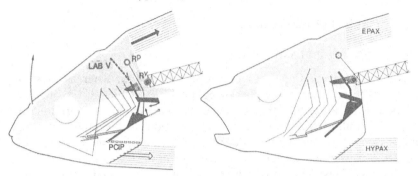

Fig. 10.   The masticatory mechanism of carp. Body muscles provide the power in crushing and grinding (indirect masticatory muscles). Epaxial muscles rotate the skull dorsad around R^v (the centre of rotation between the skull and the vertebral column), and thereby lift the pharyngeal jaws and teeth to the chewing pad through the interposed tendinous levator muscles (LAB V). The chewing pad moves mainly rostrad (black centre of rotation, black arrows). Hypaxial muscles retract the pectoral girdle around R^p (the centre of rotation between the skull and the pectoral girdle). Their forces are transmitted to pharyngeal jaws and teeth by another set of tendinous muscles (PCIP) and are added in grinding. The ligament between skull and pectoral girdle obstructs pharyngeal jaw retraction dorsally, acts as a fulcrum, and effects rotation with the jaws as long levers (open centres of rotation, open arrows). As a combined result, teeth and chewing pad move in parallel, but in opposite directions, and wedge the food under high compression. Grinding movements are small but powerful. The extensive movements of the pharyngeal jaw lever are permitted by simultaneous expansion of the buccopharynx and a slide-coupling in the branchial floor. Direct masticatory muscles that suspend the pharyngeal jaws in muscular slings, and steer and stabilise these movements, are omitted in this scheme. EPAX, epaxial muscles; HYPAX, hypaxial muscles. (From Sibbing, 1982.)

stretch the LAB V muscles thereby increasing its force production substantially (Hill, 1970). Similarly hypaxial body muscles increase the masticatory force by effectively retracting the pectoral girdle and the origin of the tendinous PCIP muscles (Fig. 10). This reinforces pharyngeal jaw rotation around R^4 and thus intensifies grinding of the teeth across the chewing pad. Electromyograms show that this chain of ventral muscles is merely active during grinding (Fig. 11). As a consequence, they expand the oropharyngeal cavity to such an extent that even the mouth is opened, which has nothing to do with food intake.

Control of mastication

Steering and stabilisation

Body muscles act indirectly in powering the system, whereas most of those directly attached to the pharyngeal jaws rather steer and stabilise. It is especially evident that the masticatory construction in cyprinid fish, where the skull provides an anvil for mastication, is developed for combining versatile movements with large forces. Such high loadings are needed to crush dried kernels of maize corn and yet the free suspension of the pharyngeal jaws demands well-steered and stabilised jaw movements. The position of the heart directly below the pharyngeal jaws makes this even more important in fish. Therefore both powerstroke muscles and their antagonists maintain their activity throughout the masticatory cycle (cf. Fig. 11), balancing the resultant loadings. The PCE thereby functions as a key antagonist counteracting the powerstroke movements around each of the axes. In addition passive elements put limits to the masticatory movements, e.g. the ligaments interconnecting the pharyngeal jaws in their symphysis and the geometrical shape of the latter, which severely restrict intrinsic jaw movements (around R^2 and R^3) in carp. Moreover, some muscles, e.g. TV V (Fig. 4b), have considerable translation components, pressing the pharyngeal jaws together in their symphysis, and so are quite analogous to the intermandibularis muscle in the rat (Beecher, 1979), thus reducing the loading of the symphysis.

Adjustment of masticatory movements to the type of food

Different types of food require different masticatory operations and, given the occlusal surfaces, changing their movements seems to be the only way to modify them. Crushing and grinding are stereotyped by the movement pattern of the pharyngeal jaws and the activity pattern of the masticatory muscles (Sibbing, 1982). Depending on the type of food and the stage of processing, a choice between these patterns is made ('switching'; Sibbing, 1991). Hard brittle materials (e.g. trout pellets or barley, Fig. 11a) are initially crushed and subsequently the small particles are ground. Soft tough food (e.g. earthworms; Fig. 11b) is immediately ground (rather squeezed) without prior crushing. Apparently the occlusal profiles of carp do not provide more effective options for soft tough items (e.g. cutting, piercing, lacerating, shearing) and in this sense carp appears to be rather specialised and limited in its 'omnivory'! Trophic specialists often have a single stereotyped neuromuscular pattern used without regard to prey consistency or size (cf. Lauder, 1983b), whereas 'a nutri-

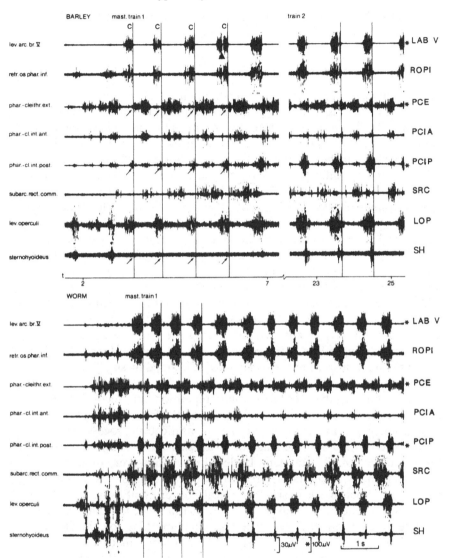

Fig. 11. In contrast to mastication of worms, mastication of barley starts, in the same experiment, with a series of crushing cycles (C). These characteristically lack activity in the ventral muscle chain rotating the pharyngeal jaws far backward in the grinding stroke (STH, PCIP), as well as the abduction effect of the PCE (see arrows). Activity patterns in grinding worms or barley are similar. The short interruptions in the LAB V and ROPI activities (see ▲) probably reflect momentary adjustments to the food. For abbreviations, see Figs. 3 and 4. (From Sibbing, 1982.)

tional jack-of-all-trades' (Liem, 1984) tends to have more versatility in the activity patterns.

Mastication as a composite process

The effectiveness of mastication depends on its component elements: food supply, loading it on to the occlusal surfaces, fracture, and clearing the teeth. Unlike mammals, fish have no movable muscular tongue for manipulation. Postcapture food transport in fish generally proceeds by teeth on the branchial arches. In cyprinids, however, food transport is effected by peristaltic waves between the muscular palatal (roof) and postlingual (floor) organ, which together propel the food like a piston (Fig. 6, frames 81–115; Sibbing et al., 1986). Similar waves load the food on to the teeth after their depression and abduction in the preparatory stroke, making way for the food. Large food items may be fixed between these organs at the entrance of the chewing cavity allowing parts to be crushed off for grinding (cf. Fig. 6, frame 272). During mastication the mucous and contractile tissue composing the walls of the chewing cavity and peculiar lateral tissue flaps will aid retention of the mucus-bound food particles on the occlusal surfaces (Sibbing & Uribe, 1985). Swallowing or deglutition proceeds by compressing the chewing cavity, at the same time closing off its entrance by the palatal and postlingual organs, thus directing transport to the esophagus. It is aided by small pharyngeal jaw movements.

Regulation of the sequence and repetition of feeding actions ('Sequence regulation'; Sibbing, 1988) is essential for increasing the feeding efficiency. It is evident that steering the masticatory process according to the type of food and the stage of processing requires information on the actual state of the food item. Little is known, however, about sensors in fish and the parameters they record. How, for example, is deglutition triggered? What particle size is critical for efficient transition from mechanical to chemical breakdown? How do fish relate this critical size to food texture? Such decisions will play an important role in efficient utilisation of food. Although taste buds may have mechanoreceptive functions and abound up to 820 mm^{-2} in the oropharyngeal cavity of carp (Sibbing & Uribe, 1985), they are almost lacking in its chewing cavity. Proprioceptors in the pharyngeal jaw muscles are expected for controlling the masticatory sequence. However, they have not been found in these muscles of carp (de Graaf & Ballintijn, 1987).

Mechanical efficiency versus rate of mastication

The metabolic rate of the animal should at least be matched by the rate of net energy gain expressed as the number of Joules gained per unit of handling time. This rate of net energy gain in feeding depends on characteristics of both prey and predator. The energy content of food will be high enough to make processing of a range of suboptimal food types still profitable. It is hypothesised that, in abundance of appropriate food, saving time has priority over saving energy. Eventually fast suboptimal processing makes more profit than maximising the energy gain at high costs of time. Only scarcity of food resources and competition will put high selective pressure on mechanical efficiency and eventually mould the feeding apparatus for increasing it for certain food types (cf. pp. 82–5).

Since food handling is a composite process, its rate is largely determined by its most time-consuming component. In predaceous fish, the search for prey and its pursuit may be the most time-consuming actions; in plankton-feeding fish, prey collection and sieving take most handling time. In molluscivorous species the ejection of shell fragments (Hoogerhoud, 1987), and in seed eating or molluscivorous species mastication, dominates handling time. Mastication evidently consumes most time (several minutes for barley) of the whole feeding sequence in carp (Fig. 9).

Masticatory efficiency can be increased by reducing the energy (by increasing the mechanical efficiency) or time (by speeding up the processing rate) spent in creating a unit of digestible food. Growth and reproduction in abundance of appropriate food will impose a premium more on the rate of food fracture than on mechanical efficiency. The rate of food fracture and the gain of digestible area per unit of time depend on the number of chews per unit of time (chewing rate), the volume broken down in each cycle (varying with number and size of teeth), and the extent of size reduction in each chew, depending on the type of comminution machinery (cf. Lucas & Luke, 1984).

Investigation of damage to chewed food and analysis of its particle size distribution over time are needed for quantifying food breakdown achievements (Lucas & Luke, 1984). Such measurements are crucial in comparing mechanical efficiencies and rates of food fracture with regard to optimal foraging, but are still completely lacking for most animals including fish.

Adapting the pharyngeal jaw system to different food size and texture

Changing the masticatory abilities according to the demands of a different food type can be achieved by modifying the construction (size, shape and position of the involved muscles, bony elements and teeth), by changing the neuromuscular activity patterns or both. A comparison of pharyngeal mastication in cyprinids with extremely different diets could show us which solutions these fish have developed for specific sizes and textures of food. It should, however, be emphasised that adaptation needs time and that environmental factors can change very rapidly. Such historical constraints or phylogenetic inertia may delay an optimal fit between demands and construction, especially in less stable environments (Liem & Wake, 1985).

Pharyngeal tooth profiles of some typical cyprinids are shown in Fig. 12. Roach (*Rutilus rutilus*) is a cosmopolitan fish, even more 'omnivorous' than common carp (Lammens & Hoogenboezem, 1991). Both fish have a markedly heterodont dentition (Fig. 12a and b). As we have seen carp (Fig. 12b) is able to crush and grind, thus fracturing hard brittle or turgid food items. Even though they can take particles up to 7% of their body length (maximal mouth diameter), only items up to 4% of their body length can enter the chewing cavity. Its actual diet (shelled molluscs, plant seeds, zooplankton and chironomid larvae) matches such categories. Roach (Fig. 12a) shows an even wider range of masticatory abilities. Whereas its front teeth may crush and pierce, its posterior teeth have sharp edges for splitting and cutting and conspicuous hooks for lacerating. Some interdigitation of opposite teeth effects shearing. This repertoire excludes small hard brittle or turgid particles, but allows most other aquatic food types to be fractured. The actual diet of roach (Table 2) is the most diverse and includes a variety of animal and plant tissues (Prejs, 1984). Grasscarp (*Ctenopharyngodon idella*) feeds preferentially on soft aquatic weeds (Hickling, 1966). Larger fish (>35 cm standard length) also feed on tough leaves of littoral plants and fibrous grasses. This fish has teeth (Fig. 12c) which alternate in position on left and right pharyngeal jaw (tooth formula 2 : 5 – 4 : 2) and they closely lock upon interdigitation through adduction of their firm pharyngeal jaws. This

---

Fig. 12.   Pharyngeal teeth of different trophic types of cyprinid fish of about 10 cm body length. (a) Pharyngeal jaws and chewing pad of roach. Only right pharyngeal teeth of (b) carp, (c) grasscarp, (d) asp and (e) silvercarp are shown in these scanning electron micrographs. Bars represent 1 mm. For further explanation, see the text.

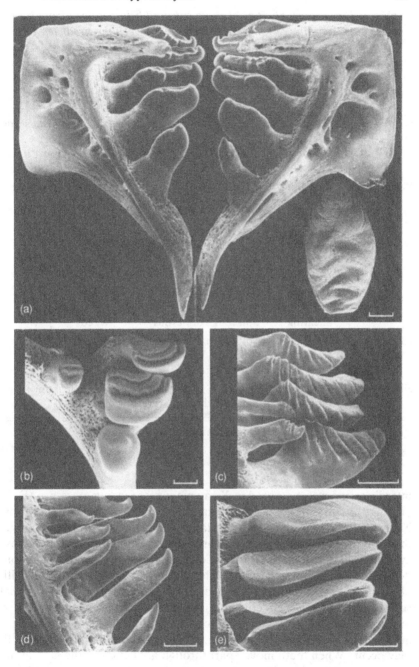

Table 2. *Masticatory abilities of some cyprinid fish, deduced from their architecture and functioning: a comparison between the predicted food range and the primary components of the actual diet*

Species	Mast. operations	Predicted food range	Actual diet
Carp	Crushing (Piercing)	Large hard brittle, turgid	Molluscs, plant seeds
	Grinding	Small hard brittle, turgid	Zooplankton, insect larvae
Roach	(Crushing) Splitting (Piercing)	Large hard brittle	Molluscs
	Cutting	Soft tough or fibrous	Insect larvae
	Lacerating (Shearing)	Tough fibrous	Weeds
Grasscarp	(Grinding) Cutting	Soft tough or fibrous	Weeds
	Lacerating	Tough fibrous	Weeds
	Shearing	Soft tough or fibrous	Weeds
Asp	Piercing (Cutting)	Soft tough	Fish
	Lacerating	Tough fibrous	Insects, fish
Silvercarp	(Crushing)		Phytoplankton
	Grinding	Small hard brittle, turgid	

Diet data from Lammens & Hoogenboezem (1991).

interdigitation was measured in X-ray movies of feeding grasscarp (F. A. Sibbing, unpublished data) to overlap about 30% of their toothcrowns. X-ray analysis showed little skull rotation but extensive transverse movements of the pharyngeal jaws compared to carp. Unlike carp this Chinese cyprinid is almost unable to crush and grind. Grasscarp have a homodont dentition (Fig. 12c). Each tooth is laterally compressed and provided with oblique ribs at its anterior and posterior surface, giving the occlusal surface serrated edges effective in cutting against the chewing pad. When in the preparatory stroke teeth of both sides interdigitate and food items are brought upon their occlusal area, shearing between interlocking teeth may occur. When teeth move apart through abduction of the jaws in the power stroke, food is cut and lacerated against the horny chewing pad, which is even more sculptured by ridges and furrows than in roach (Fig. 12a). Such masticatory abilities might explain the common use of these fish as complements in polyculture (Bardach, Ryther & McLarney, 1972) to exploit all available food resources. Fragments of aquatic weeds collected from their faeces are of about the same size as the spaces between adjacent teeth (Hickling, 1966; Vincent & Sibbing, 1991). Similar results were obtained for rudd (*Scardinius erythrophtalmus*) feeding on brittle macrophytes (Prejs, 1984). Pharyngeal teeth of rudd also have serrated cutting edges (Sibbing, 1984) and closely resemble those of grasscarp. Both the pharyngeal jaws and their levators and abductor muscles (LAB V) are of impressive size in grasscarp compared to other species, reflecting the high power and loading of the system in this macroherbivore. However, microdiminution is very poor and only cells at the cut edges are ruptured and their contents digested. This renders the creation of cut edges through macrodiminution even more crucial in feeding.

The asp (*Aspius aspius*), a typical carnivorous cyprinid, has slender pharyngeal jaws each bearing slender medially tapering teeth (Fig. 12d), not interdigitating from opposite sides (tooth formula 3 : 5 – 5 : 3). Each tooth has a cutting edge and a hook, both not very conspicuous. Such teeth allow mainly piercing during adduction, and lacerating during abduction of the jaws in the powerstroke. Piercing allows entrance of digestive enzymes into the food but, more importantly, also provides grip for transporting such large prey as fish towards the oesophagus. In the meantime the side of the fish exposed to the teeth is lacerated or tough insect cuticles are torn into pieces. The silver carp (*Hypophtalmichthys molitrix*) is another Chinese specialist kept in polyculture with carp and grasscarp as it exploits only the minute phytoplankton. Its flat spatulate teeth (Fig. 12e; tooth formula 4–4) show a delicate parallel pattern of microridges on the surface. Whereas grasscarp effects mainly macrodiminution of weeds, silver carp focuses on microdiminution of algae. The sideways implantation of these teeth and the most delicate architecture of their pharyngeal jaws hardly permit crushing. This outfit is evidently one of the most specialised for grinding thus far encountered. Investigation of the faeces of this fish (Kajak, Spodniewska & Wisniewski, 1977) again shows most algae undamaged, once more stressing the high demands on feeding rate in herbivorous fish. However, pharyngeal mastication does extend the exploitation of aquatic food resources to such otherwise (for cyprinids) unutilisable food items, which are continuously abundant in tropical areas. *Hypophtalmichthys nobilis*, the bighead, has almost identical pharyngeal jaws and teeth compared to the silver carp, but lacks microridges on its tooth crowns. Bigheads feed on zooplankton exclusively and squeeze them with their smooth spatulate teeth against the chewing pad. Bighead is another component of Chinese polyculture systems, in which complementary trophic specialists are kept to exploit all levels of the available resources maximally for fish growth.

Direction and intensities of masticatory movements differ widely among these cyprinids: in carp, vertical crushing movements dominate; in grasscarp, it is transverse lacerating and in asp forward and backward movements (F. A. Sibbing, unpublished data). Neuromuscular patterns hardly change. This imposes different demands on the driving machinery (position of teeth and rotational centres, length of levers, position, direction, length and cross-sectional area of muscles). The supporting architecture and dimensions of the pharyngeal jaws clearly reflect the size and direction of loadings which are reacting on their teeth. Relations between pharyngeal jaws and food texture also become apparent in the study of the growth of fish larvae. The pharyngeal dentition in a young cyprinid larva rapidly changes by developing new tooth generations. The changes in shape and number of teeth until the juvenile stage are at least partly related to the rapidly changing diet during larval growth.

Mastication and ecological types

The feeding niche is not merely dictated by masticatory abilities, since specialisations in other feeding actions (Fig. 1) limit the utilisable foods as well (Sibbing, 1988). Accumulation of such limitations will narrow the feeding niche severely. Figure 13 shows the extremes of different food modalities. It is assumed that for efficient processing, specialisations for such extremes are not compatible. The fish's intestine must specialise either for animal or for plant food digestion. Among animal food, mastication can specialise on hard brittle or soft and ductile foods. Subsequently, specialisation for prey capture depends largely on prey velocity and size, whereas prey position (benthic versus pelagic) demands further specialisation (e.g. fleshy sucking mouths in carp and bream). If benthic food is heavily mixed with non-food their oral separation requires a subtle sorting device such as the palatal organ in cyprinids. In general, food niches become narrow with increasing specialisation (but see Liem, 1984). Generalists will have more intermediate positions for handling one or more food modalities. Only a few food properties (e.g. chemical composition) are strictly bound to feeding guilds as herbivores and carnivores. Therefore a definition of feeding types based on a differentiated pattern of food handling is more appropriate. A thorough analysis of the architecture and functioning of the total feeding apparatus and an evaluation of its body shape for prey pursuit (Webb, 1982) can thus predict the feeding niche (Sibbing, 1988) and the ecological type.

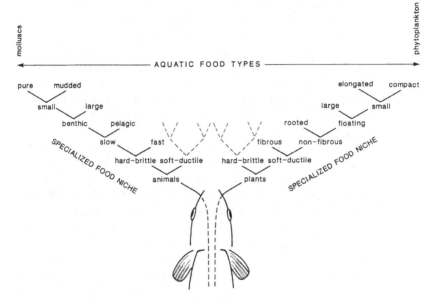

Fig. 13. Limitation of utilisable food types starting from the fish's intestine. Each food type has an array of characters linked to specific subactions in food handling. Specialisations and consequent limitations may develop for each of these characters. As a result, a large set of ultimate specialists is possible; for example, the one depicted left in this figure specialised on a very narrow food niche (e.g. benthic molluscs) and the one at the right specialised on phytoplankton. Less specialisation in one or more characters widens the food niche and may gradually lead to generalists with a high trophic versatility.

Conclusions

1. Mastication is not common among fish. Oral jaws serve to seize and bite prey and, primitively, buccopharyngeal teeth serve transport and swallowing. True mastication is bound to the evolutionary development of posterior branchial arches into pharyngeal jaws as in some lower (e.g. cyprinid) and many higher (e.g. cichlid) teleosts. They are freely suspended in muscular slings from the skull.

2. Cyprinid fish lack oral teeth, a stomach and cellulases. High demands are thus imposed on pharyngeal mastication for increasing the surface area of food for effective digestion or for breaking indigestible capsules. The movably connected

pharyngeal jaws oppose a horny pad far back in the skull and are designed for transmitting high forces.

Most other fish have oral teeth, a stomach, and some of them cellulases. In addition cichlid fish, for example, independently evolved fused lower pharyngeal jaws working as a single unit against upper pharyngeal jaws, which articulate with the posterior skull. Compared with cyprinids, they probably combine less high forces with a greater versatility in movement.

3. In comparison to mammals, fish have no claws or movable tongue for the manipulation of food and have to cope with water taken in with the food. Apart from mastication, the oropharyngeal complex has many other functions including respiration, suction of food, sieving and purification. Oral jaws in mammals function more independently, combine biting and mastication, have less versatility in movements and will achieve higher loadings on the food. As a result masticatory machineries in fish and mammals are difficult to compare.

4. The type of loading on the food depends on the size, shape and movement of opposing occlusal surfaces. Mechanical properties of aquatic foods and insight into how they break are poorly known and urgently need further investigation. For the present purpose masticatory operations are defined (crushing, splitting, piercing, grinding, cutting, lacerating and shearing) and food is subdivided into four main classes (hard brittle, turgid, soft tough, tough fibrous).

5. Mastication in cyprinids is a lasting cyclic event. In carp both crushing and grinding cycles are distinct, each stereotyped in its neuromuscular and movement pattern and employed according to the type of food and the stage of processing. Its heterodont teeth thus permit effective macro- and microdiminution of hard, brittle and turgid materials which is matched by its actual diet. The size of the chewing cavity limits items to be masticated to a diameter of about 4% of the body length in carp.

6. The pharyngeal jaw system in carp is built for transmitting and absorbing high forces. Body muscles power the pharyngeal jaw system, whereas most pharyngeal jaw muscles largely steer and stabilise the masticatory movements.

7. Preliminary comparison of trophic extremes among cyprinids (roach, grasscarp, asp, silvercarp) show that adaptation to

different food size and texture is based more on different dentitions than on major changes in neuromuscular patterns.

8. Masticatory abilities do not dictate the food niche of fish. However, they extend the exploitation of aquatic food resources to plant materials and hard items (e.g. snails) previously little utilised by fish. The eventual food niche is narrowed by specialisations and consequent limitations accumulated from each separate feeding action. The definition of feeding types on the mode of prey handling is more appropriate than distinction between herbivores, carnivores and omnivores.

9. Mastication takes much handling time. It is hypothesised that with abundant food, increasing the rate of food breakdown (creation of new digestible area per unit of time) has more priority for growth and reproduction than maximising the mechanical efficiency at high costs of time. Measurements of the degree of break down of food need to be further developed to test such a hypothesis.

References

Alexander, R.McN. (1983). *Animal Mechanics*. Blackwell Scientific Publications, London.

Ballintijn, C.M. (1969). Functional anatomy and movement coordination of the respiratory pump of the carp (*Cyprinus carpio* L.). *Journal of Experimental Biology*, **50**, 547–67.

Bardach, J.E., Ryther, J.H. & McLarney, W.D. (1972). *Aquaculture*. Wiley & Sons Inc., New York.

Barel, C.D.N. (1983). Towards a constructional morphology of cichlid fishes (Teleostei, Perciformes). *Netherlands Journal of Zoology*, **33**, 357–424.

Barel, C.D.N., van Oijen, M.J.P., Witte, F. & Witte-Maas, E. (1977). An introduction to the taxonomy and morphology of the haplochromine Cichlidae from Lake Victoria. *Netherlands Journal of Zoology*, **27**, 333–89.

Beecher, R.M. (1979). Functional significance of the mandibular symphysis. *Journal of Morphology*, **159**, 117–30.

Castro, N.M., Sasso, W.S. & Katchburian, E. (1961). A histological and histochemical study of the gizzard of the mugil sp. *Pisces* (Tainha). *Acta Anatomica*, **45**, 155–63.

Chu, Y.T. (1935). Comparative studies on the scales and on the pharyngeals and their teeth in Chinese cyprinids, with particular reference to taxonomy and evolution. *Biological Bulletin, St John's University Shanghai*, **2**, 1–225.

Daget, J. (1962). Le genre *Citharinus* (Characiformes). *Revue Zoologica et Botanica Africana*, **61**, 81–106.

de Graaf, P.J.F. & Ballintijn, C.M. (1987). Mechanoreceptor activity in the gills of the carp. II. Gill arch proprioceptors. *Respiration Physiology*, **69**, 183–94.

Fischer, Z. (1968). Food selection in grasscarp (*Ctenopharyngodon idella* Val.) under experimental conditions. *Polskie Archives Hydrobiologica*, **15**, 1–8.

Freyer, G. & Iles, T.D. (1972). *The Cichlid Fishes of the Great Lakes of Africa: Their Biology and Evolution*. Oliver and Boyd, Edinburgh.

Gordon, J.E. (1976). *The New Science of Strong Materials*. Penguin, Harmondsworth, Middx.

Goulding, M. (1985). Forest fishes of the Amazon. In *Key Environments of Amazonia*, ed. G.T. Prance & T.E. Lovejoy, pp. 267–76. Pergamon Press, New York.

Gregory, W.K. (1933). Fish skulls. A study of the evolution of natural mechanisms. *Transactions of the American Philosophical Society*, **23**, 475–81.

Heckel, J.J. (1843). Abbildungen und Beschreibungen der Fische Syriens nebst einer neuen Classification und Characteristik sammtlicher Gattungen der Cyprinen. *E. Schweizerbart'sche Verlagshanlung*, Stuttgart.

Hickling, C.F. (1966). On the feeding process in the White Amur. *Ctenopharyngodon idella. Journal of Zoology, London*, **148**, 408–19.

Hiimae, K.M. & Crompton, A.W. (1985). Mastication, food transport and swallowing. In *Functional Vertebrate Morphology*, ed. M. Hildebrand, D.M. Bramble, K.F. Liem & D.B. Wake, pp. 262–90. Harvard University Press, London.

Hill, A.V. (1970). *First and Last Experiments in Muscle Mechanics*. Cambridge University Press, Cambridge.

Hofer, R. (1991). Digestion. In *The Biology of Cyprinids*, ed. J. Nelson & I.J. Winfield. Chapman and Hall, London, in press.

Hoogenboezem, W., Sibbing, F.A., Osse, J.W.M., van den Boogaart, J.G.M., Lammens, E.H.R.R. & Terlouw, A. (1990). X-ray measurements of gill arch movements in filterfeeding bream (*Abramis brama*, Cyprinidae). *Journal of Fish Biology*, **36**, 47–58.

Hoogerhoud, R.J.C. (1987). The adverse effects of shell ingestion of molluscivorous cichlids, a constructional morphological approach. *Netherlands Journal of Zoology*, **37**, 277–300.

Hoogerhoud, R.J.C. & Barel, C.D.N. (1978). Integrated morphological adaptations in piscivorous and mollusc-crushing *Haplochromis* species. In *Proceedings of the Zodiac Symposium on Adaptation*, pp. 52–6. Centre for Agricultural Publishing and Documentation, Wageningen.

Kajak, Z., Spodniewska, I. & Wisniewski, R.J. (1977). Studies on food selectivity of silver carp, *Hypophtalmichthys molitrix* (Val.). *Ekologia Polska*, **25**, 227–39.

Lammens, E. H. R. R. & Hoogenboezem, W. (1991). Diets and feeding behaviour. In *The Biology of Cyprinids*, ed. J. Nelson & I. J. Winfield. Chapman and Hall, London, in press.

Lauder, G. V. (1983*a*). Functional design and evolution of the pharyngeal jaw apparatus in euteleostean fishes. *Zoological Journal of the Linnean Society, London*, **77**, 1–38.

Lauder, G. V. (1983*b*). Functional and morphological bases of trophic specialization in sunfishes (Teleostei, Centrarchidae). *Journal of Morphology*, **178**, 1–21.

Leeuwen, J. L., van & Muller, M. (1984). Optimum sucking techniques for predatory fish. *Transactions of the Zoological Society of London*, **37**, 137–69.

Liem, K. F. (1973). Evolutionary strategies and morphological innovations: cichlid pharyngeal jaws. *Systematic Zoology*, **22**, 425–41.

Liem, K. F. (1984). Functional versatility, speciation and niche overlap: are fishes different? In *Trophic Interactions within Aquatic Ecosystems*, Selected Symposium 85, ed. D. G. Meyers & J. R. Strickler, pp. 269–305. American Association for the Advancement of Science, Washington, DC.

Liem, K. F. & Greenwood, P. H. (1981). A functional approach to the phylogeny of the pharyngognath teleosts. *American Zoologist*, **21**, 83–101.

Liem, K. F. & Wake, D. B. (1985). Morphology: current approaches and concepts. In *Functional Vertebrate Morphology*, ed. M. Hildebrand, D. M. Bramble, K. F. Liem & D. B. Wake, pp. 366–77. Harvard University Press, London.

Lucas, P. W. (1979). The dental–dietary adaptations of mammals. *Neues Jahrbuch der Geologie und Paläontologie*, **8**, 486–512.

Lucas, P. W. (1982). Basic principles of tooth design. In *Teeth: Form, Function and Evolution*, ed. B. Kurten, pp. 154–62. Columbia University Press, New York.

Lucas, P. W. & Luke, D. A. (1984). Chewing it over: basic principles of food breakdown. In *Food Acquisition and Processing in Primates*, ed. D. J. Chivers, B. A. Wood & A. Bilsborough, pp. 283–301. Plenum Press, New York.

Nelson, G. J. (1969). Gill arches and the phylogeny of fishes, with notes on the classification of vertebrates. *Bulletin of the American Museum of Natural History, New York*, **141**, 475–552.

Nelson, J. S. (1982). *Fishes of the World*. John Wiley & Sons, New York.

Nikolsky, G. V. (1963). *The Ecology of Fishes*. Academic Press, London.

Prejs, A. (1984). Herbivory by temperate freshwater fishes and its consequences. *Environmental Biology of Fishes*, **10**, 281–96.

Rensberger, J.M. (1973). An occlusion model for mastication and dental wear in herbivorous mammals. *Journal of Palaeontology*, **47**, 515–28.

Sibbing, F.A. (1982). Pharyngeal mastication and food transport in the carp (*Cyprinus carpio* L.): a cineradiographic and electromyographic study. *Journal of Morphology*, **172**, 223–58.

Sibbing, F.A. (1984). Food handling and mastication in the carp. Ph.D. thesis, Agricultural University, Wageningen.

Sibbing, F.A. (1988). Specializations and limitations in the utilization of food resources by the carp, *Cyprinus carpio*: a study of oral food processing. *Environmental Biology of Fishes*, **22**, 161–78.

Sibbing, F.A. (1991). Food capture and oral processing. In *The Biology of Cyprinids*, ed. J. Nelson & I.J. Winfield, pp. 377–412. Chapman and Hall, London.

Sibbing, F.A., Osse, J.W.M. & Terlouw, A. (1986). Food handling in the carp (*Cyprinus carpio* L.), its movement patterns, mechanisms and limitations. *Journal of Zoology, London*, series A, **210**, 161–203.

Sibbing, F.A. & Uribe, R. (1985). Regional specializations of the oropharyngeal wall and food processing in the carp (*Cyprinus carpio* L.). *Netherlands Journal of Zoology*, **35**, 377–422.

Stein, R.A., Kitchell, J.F. & Knezević, B. (1975). Selective predation by carp (*Cyprinus carpio* L.) on benthic molluscs in Skadar Lake, Yugoslavia. *Journal of Fish Biology*, **7**, 391–9.

Vasnecov, V. (1939). Evolution of pharyngeal teeth in Cyprinidae. In *A la Mémoire de A.N. Sewertzoff*, pp. 439–91. Academy of Sciences USSR, Moscow.

Vincent, J.F.V. & Sibbing, F.A. (1991). How the grasscarp (*Ctenopharyngodon idella*) chooses and chews its food – some clues. *Journal of Zoology, London* (in press).

Webb, P.W. (1982). Locomotor patterns in the evolution of actinopterygian fishes. *American Zoologist*, **22**, 329–42.

Werner, E.E. & Hall, D.J. (1974). Optimal foraging and the size selection of prey by the bluegill sunfish (*Lepomis macrochirus*). *Ecology*, **55**, 1042–52.

Zander, E. (1906). Das Kiemenfilter der Teleosteer, eine Morphophysiologische Studie. *Zeitschrift für Wissenschaftliche Zoologie*, **84**, 619–713.

P. W. LUCAS AND R. T. CORLETT

Quantitative aspects of the relationship between dentitions and diets

The study of the systematic size reduction of solids poses special problems for mechanical analysis because the continually changing geometry of the loading between working surfaces of the equipment and the solid to be comminuted renders any exact stress analysis impossible. Even given that a stress pattern could be established, it is not at present possible to model stresses beyond the point at which a discontinuity (crack) starts to propagate in the solid, let alone the multiple and repetitive fracture which defines comminution; the shape and position of cracks and particle fragments cannot be specified. The chewing process is usually a comminution process and is additionally complicated by being masked by the cheeks. However, given sufficient information about the food input and the swallowed output coupled with static observations of the teeth in and out of the oral cavity of mammals, some aspects of the process become understandable. Given abundant information on dental morphology, the missing factor in the analysis of dentitions and diets is the almost total absence of knowledge of food texture.

This chapter concentrates more on food than teeth and emphasises work on primates because the behaviour and diet of these mammals has been studied intensively. Primates inhabit the tropics and subtropics forming a significant proportion of the arboreal animal biomass in primary rain forest (Terborgh, 1983). The vast number of animal and plant species in rain forest (Whitmore, 1984) affords the opportunity to study dietary selectivity, but with the drawback that complexity increases more than in proportion to the number of species in the environment because of the possibility of networks of coevolved plant–animal relationships. Primate diets range from small vertebrates and invertebrates to the structural and reproductive parts of plants.

Our understanding of oral processing in humans is severely limited by the difficulty of making direct measurements. Suggested animal models include pigs (Herring, 1976) and monkeys, in particular, *Macaca fascicularis*, the long-tailed macaque (Byrd, Milberg & Luschei, 1978). This

Southeast Asian Old World monkey is the most common laboratory monkey. Though substantially different in its oral anatomy and physiology from man (Hylander, Johnson & Crompton, 1987), continual reference will be made to its feeding behaviour and diet because we have studied it intensively in Bukit Timah Nature Reserve, a small patch of largely primary rain forest in Singapore (Corlett, 1988, 1990). There is a further emphasis on fruits and seeds because these form a large part of the diet of primates and many other mammals. Vertebrate tissues and other plant foods are covered elsewhere in this volume.

Method of analysis

Previous views

Without consideration of what a structure does, nothing can be said of its function. If it is accepted that the function of teeth is to fracture food and usually to reduce its particle size prior to swallowing, then knowledge of the relevant properties of foods and teeth is a prerequisite for theories that attempt to predict tooth designs for particular diets. However, in the mammalian literature, this has not been a common view. Instead, a type of analysis which makes no overt reference to food properties has been adopted. Simpson (1936) expresses this clearly: 'the fact is that mammals use their teeth to acquire and prepare their food and that their molars are in this sense adapted to certain types of activity and beyond this to certain types of diet'. By interpolating subcategories of fracture activity between teeth and foods, a 'functional' analysis of teeth can be devised that refers only to activity and not to food. Consequently, work on food has been either incidental or unnecessary. The stultifying effect of this activity terminology has been extreme.

The activities listed by Simpson and many others are essentially crushing, shearing (equal to cutting for many authors) and grinding. According to different authors the molar dentition of man shears (Shaw, 1917), crushes (Lumsden & Osborn, 1977) or grinds (Waters, 1980). Since this covers the entire range, it is difficult to understand why our dentition is so different from that of the cat or cow. These familiar words, which have a history stretching from (translations of) Aristotle (Peck, 1937) to a recent textbook (Hiiemae & Crompton (1985), who include chomping), connote something unspecific about the movement of surfaces and are part of a rich vocabulary of over a hundred words in English. Many of these words appear sporadically in the literature. However, all refer to successful events in that fracture of an object between these surfaces is accomplished. The terms are therefore incapable of analysing the conditions

Table 1. *Classification of textural
characteristics*

External properties	Internal properties
Size	Toughness
Shape	Deformability
Total volume	Strength
Roughness	
Stickiness	
Abrasiveness[a]	

[a]Measureable only over a time span.

necessary for fracture to take place. They are perfect for eulogies about
the efficiency of fossil dentitions.

Classification of food texture

Classifications have been offered by Szczesniak (1963) and Sherman
(1969). Essentially geometrical and mechanical properties have been
separated in varying ways. The classification of Lucas, Corlett & Luke
(1986a), used here, differs in that properties that act at the food surface
(and therefore interact with tooth surfaces) are distinguished from those
properties resisting the formation of new surface (Table 1). External
properties include the size, shape and total volume of particles in the
mouth, stickiness (adhesion), roughness but also abrasiveness. By
jeopardising the ability of the teeth to continue to operate, abrasiveness
can be considered as a surface property whose effects can only be
measured over a time span. Internal properties include modulus of elasticity, strength and a measure of fracture toughness.

Stages of oral food processing

The oral processing of foods, whether liquid or solid, involves subjecting
them to forces that may move or break them. Initially, food is ingested in
the mouth by a highly variable procedure often involving fracture. The
food is then manoeuvred towards the premolar and molar teeth where it
is masticated. Following this, it is again transported to a site where it is
accumulated prior to swallowing. The transport phases pre-date mammalian evolution (Hiiemae & Crompton, 1985) and are significant particularly in long-jawed animals. The major activity that distinguishes

mammals and determines the length of time that food is resident in the mouth is mastication. Mammals are probably very sensitive to food texture; certainly the oral or circum-oral receptors that receive information from which textural judgments are made in man are found and investigated in other mammals (reviewed by Heath & Lucas, 1988).

Only 'solid' foods, which preserve their shape under short gravitational loads (some fraction of the duration of a chewing cycle), can be chewed because only these can be gripped. Low-amplitude jaw movements are associated with the intra-oral transport of liquid foods but this is probably a corollary of tongue movement (Hiiemae & Crompton, 1985). The objectives of oral processing are:

1. To make foods that are impervious to the chemical action of the gut digestible. This is particularly important for herbivores (Macarthur & Sanson, 1988; Vincent, 1990). Lanyon & Sanson (1986) show that koalas swallow much larger food particles when their teeth are very worn.
2. To increase the speed of action of gut enzymes. This is the only general explanation for the oral comminution of foods by mammals.
3. To make foods swallowable. This is particularly important for carnivores that swallow large particles.

Each of these objectives can be helped or met by reducing the particle size of solid foods.

Characterisation of the chewing process

The purpose of the chewing process could be stated as a directed change in the external properties of foods. When this change results in a reduction in food particle size, which is usual for mammalian foods, two aspects of comminution can be analytically separated. Any particle in the mouth has a probability of being fractured and, upon fracture, a distribution of fragment sizes. These two aspects have been defined as a selection function and a breakage (or distribution) function, respectively, and measured in human mastication (Lucas & Luke, 1983a; Olthoff, 1986) with rather more ease than in the industrial processes for which they were intended. This is because food particles can be shaped and coloured according to size prior to offering them for chewing which allows identification of parent particles and their fragmented offspring. For one food (carrot – Lucas & Luke, 1983a, b) and a chewable material (elastomeric silicone rubber – Olthoff, (1986)), the selection function has been related to food particle size by a power law with an exponent of between 1.5 and

3.0. Small particles have a very much lower probability of fracture than large ones. Computer modelling of empirically derived values for selection and breakage functions (Lucas & Luke, 1983b; Olthoff, 1986; Voon *et al.*, 1986) shows some limited success in predicting particle size distributions for these brittle materials.

Increase in the volume of food in the mouth (called the mouthful) reduces the rate of particle size reduction but produces particle size distributions that are geometrically similar. This suggests that the selection function varies strongly with the mouthful but the breakage function does not (Lucas & Luke, 1984). An increase in the mouthful may decrease the rate of particle size reduction but it also increases the rate at which new food surface is exposed (Lucas & Luke, 1984). This increase must have some limit because humans at least choose mouthfuls far below the actual capacity of the oral cavity to accommodate food. There is evidence that much of the tooth row on the working side of the mouth is not utilised to break food in any one chew (Yurkstas & Curby, 1953; Wictorin, Hedegard & Lundberg, 1968; Lucas & Luke, 1984; Tornberg *et al.*, 1985). It is therefore necessary to postulate extensive circulation of particles in the mouth by the tongue to account for unimodal particle size distributions. These points are relevant to the discussion of wear below.

Food preferences

The immediate decision by a mammal, as observed in a forest, about whether to ingest a food item could involve its availability. This needs to be taken into account before preferences can be assessed. Availability of the mature structural parts of plants is continuous. However, most of the plant biomass is woody and therefore inedible due to its lignification. Many primates have diets that vary strongly with time of year (Milton, 1980; van Roosmalen, 1980; Terborgh, 1983), showing that those mature plant parts from which some sustenance can be obtained (mostly mature leaves and bark), are relatively unimportant.

In order to ascertain preference for particular foods, it is necessary to compare the availability of potential food items with consumption patterns. It is not possible to separate the data in terms of physical or chemical attributes on present knowledge and so we start by distinguishing fruit, flowers, leaves and animal matter. Our own study (P. W. Lucas & R. T. Corlett, unpublished data) on *M. fascicularis* was for a 68 week period and showed very strong correlations between the availability of fruit and its consumption on a weekly basis. This was true whether measured in terms of number of plant species fruiting and consumed or individual trees that were fruiting compared with standardised observations of fruit-eating by the monkeys. The preference was for fruits with a moisture-laden flesh because non-fleshy fruits were underrepresented in the diet. Correlations for flowers were much lower. It is notable that the peak in fruit or flower production could be timed to the week (Corlett, 1990).

We found that foraging for insects decreased significantly when fruit consumption was raised. This would be explained by primatologists on the basis that insect capture, which is manual, is very slow and only small primates (less than 500 g body weight) can fulfil their metabolic requirements through consumption of insects (Kay, 1975; Terborgh, 1983). In tropical rain forests, insect abundance varies during the year (Murphy, 1973). Consumption of vertebrates is probably insignificant in most primates, even baboons and chimpanzees.

It could be that taste, aroma or absence of orally undetectable toxins determines fruit choice (Janzen, 1978; Harborne, 1982). Though chemical properties are thought to play the major role (Harborne, 1982; Waterman, 1984), the evaluation of toxicity in terms of the least dose that would kill 50% of animals is obviously impossible. Evidence given below suggests that texture is important and is certainly the key factor in determining the manner of oral processing.

Ingestion

Internal physical properties

Fruits either possess a moisture-laden flesh (termed fleshy) or do not (termed dry). Partition of a sample of fruit species into these categories is usually unequivocal. Though *M. fascicularis* apparently avoided dry fruits, we assume that the opposite may be true for rodents such as squirrels, which tend to concentrate either on dry fruits or ignore fruit flesh and eat seeds.

A further subdivision of fleshy fruits is possible on textural grounds by examining the nature of the outermost tissue. Janson (1983) found that most species can be easily partitioned into those fruits that have thick non-woody peels above 10% of the total fruit thickness, which he termed protected, and those that are covered by only thin flexible skins which were physically unprotected. It is preferable to specify an absolute measure because this is mechanically relevant and otherwise, for example, fruits of rattans (Palmae), which possess scaly peels of less than 5% of fruit thickness, would be classified as unprotected. However, even this is inadequate. The thin weak skin of 'obviously' unprotected *Mezzettia leptopoda* (Annonaceae) fruits cannot be cleanly removed from the ripe flesh by hand or teeth but has the same thickness as the peel of *Calamus*

Table 2. *Peel thickness of some forest fruits in relation to fruit size*

Species	Peel thickness (mm)	Fruit width (mm)	Peel as % of fruit width
Alangium nobile (Alangiaceae)	1.8	25	7
Cyathocalyx ramuliflorus (Annonaceae)	1.7	21	8
Mezzettia leptopoda (Annonaceae)	0.6	52	1
Willughbeia coriacea (Apocynaceae)	5.0	30	17
Gnetum microcarpum (Gnetaceae)	1.5	14	11
Garcinia cf. *forbesii* (Guttiferae)	2.9	27	11
Fibraurria tinctoria (Menispermaceae)	1.9	20	10
Calamus luridus (Palmae)	0.6	14	4
Xerospermum sp. (Sapindaceae)	2.2	20	11

luridus, a rattan palm (Table 2). Without quantification of mechanical properties of the outer layer of fruits (such as for cultivated fruits, Vincent, 1990), the classifications remain subjective. Of course, much more complicated schemes are possible (Whitten, 1982; Ng, 1988), but these are again texturally based.

Janson (1983) found that frugivorous birds in Amazonian forest ate unprotected fruit whereas mammals (mostly primates) ate equal numbers of protected and unprotected fruit species. The clear implication is that most frugivorous birds lack the ingestive equipment to cope with the removal of thick peels. Primates process protected fruits by holding them in the hand(s) and removing the peel with the incisors. The ability of primates to consume unprotected fruits is restricted by competition from birds whereas texture prevents birds from consuming peely fruits.

It is not possible to characterise all fruits as protected or unprotected on peel characteristics. Treating the weak outer casing of almost any fig (genus *Ficus*, Moraceae) as protected makes no mechanical sense. Dehiscent fruit can be considered as unprotected but the whiskers of *Neesia altissima* (Bombacaceae) shown in Fig. 1 deterred macaques; one monkey was observed reaching into the fruit and then spending several aggravated minutes rubbing his hands in a similar manner to humans foolish enough to do this.

The consumption of foliage is often viewed as constrained by chemical defences. However, Coley (1983) surveyed a large number of plant species in Barro Colorado and found that the force required to push a plunger through the lamina was much more highly correlated with the

Fig. 1. Left, a dehisced *Neesia altissima* fruit which, instead of offering a brightly coloured aril as reward to a potential disperser, has dull-red seeds set in a hairy bed (arrowed). Scale bar represents 10 mm. Right, some of the razor-sharp hairs which contain silicon (perhaps as silica), potassium and phosphorus (electron probe X-ray microanalysis by Dr David Lane). Scale bar represents 100 μm.

amount of damage to leaves (largely invertebrate) than was any chemical measure or nutrient content. Subsequently, Coley (1987) studied seven sites and found significant correlations with this textural measurement at all locations. Unfortunately, no fundamental significance can be attached to the measurement (even lamina thickness is not given). However, experiments on captive beetles, with leaves whose physical dimensions were relatively controlled (Raupp, 1985), showed that the beetles wore their mandibles faster when feeding on leaves that were 'tougher', and that this then slowed their ingestion rate and lowered their fecundity. This suggests that the fracture toughness of leaves is ecologically important.

Macaca fascicularis regularly ate the immature leaves of several species, the laminae of all of which appeared qualitatively to have low fracture resistance, the leaves of *Xanthophyllum maingayi* (Polygalaceae) being extreme in this respect. Almost all leaves were ingested either by bringing a small branch towards the face with a hand and biting with the incisors against the pull of the hand or by biting off with the incisors direct from the branch. Both methods were employed for *Xanthophyllum*. However, the most common leaf eaten, *Gluta wallichii* (Anacardiaceae), was ingested in a unique way. The lamina of a young leaf was always stripped off the main vein (which is noticeably stout) by hand and pushed into the mouth. The tree contains alkyl catechols (Baer, 1983), which are toxic and to which most humans are allergic on contact, but which appear not to afflict the monkeys.

External physical properties

Once a peel has been removed, when present, from a fleshy fruit, the deciding factor for mode of ingestion is its size. There is little evidence that primates select fruits on the basis of fruit size. Terborgh (1983) found primates varying in an order of magnitude in body size ate from almost, but not quite, the same fruit size range. However, whether the fruit can enter the mouth or instead be processed by the incisors from the hands must be determined by the size of the mouth-slit and oral cavity. Fruit sizes in Bukit Timah range from 4 mm in maximum width to over 110 mm. Long-tailed macaques, which have a mouth-slit up to about 35 mm wide when measured in the plane of the incisal edges, ate over the entire range. However, the largest fruits, *Artocarpus* spp. (Moraceae – protected), *Willughbeia coriacea* (Apocynaceae – protected), *Mangifera indica* (Anacardiaceae – unprotected), *Milletia atropurpurea* (Leguminosae – protected) and *Nothaphoebe umbelliflora* (Lauraceae – unprotected) did not enter the mouth most probably because they were too large.

The flesh of uncultivated fruits is generally very thin (down to 1 mm). For example, the reward for consumption of a single *Embelia ribes* (Myrsinaceae) fruit, avidly eaten by *M. fascicularis*, is 0.4 cm³ of flesh. As a consequence, in many species, the total weight of seeds is often over half the weight of the whole fruit (Whitten, 1982). Seed size and fruit size are highly correlated in single-seeded fruit but not in multi-seeded fruits (Fig. 2) and are probably best considered separately. Seed size appeared to

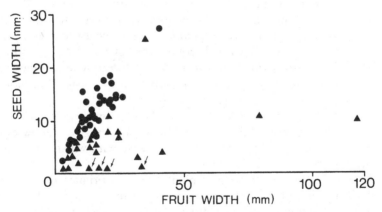

Fig. 2. The relationship between the maximum widths of 57 species (29 families) of fruits and their contained seeds. For single-seeded fruits (circles), the relationship is strong ($r=0.90$, $p<0.001$, $n=33$) but not for multi-seeded fruits (triangles) ($r=0.44$, $p<0.05$, $n=24$). Four *Ficus* species are arrowed; their seed sizes do not depend on fruit size.

govern decisions by *M. fascicularis* about whether to remove flesh from a seed with the incisors in front of the mouth or inside with the postcanines.

Stickiness is an important textural defence of plants. Latex production by members of many plant families (e.g. Sapotaceae, Anacardiaceae, Moraceae, Guttiferae and Euphorbiaceae) may deter biting insects (Janzen, 1985). The stickiness of latex is a textural defence often present in the mature fruit peel but in smaller amounts. Latex appears to fail to deter primates from ingesting fruits – e.g. *Garcinia* (Guttiferae), which is a key genus of fruit for Southeast Asian primates (Caldecott, 1986). The main leaf species in the diet of *M. fascicularis* (*Gluta wallichii* and members of the Moraceae) contained latex. It is unlikely that monkeys prefer latex (the macaques licked their lips and rubbed their faces on tree trunks after consumption of some latexy fruits) but rather that these plants may not have any other significant chemical defences.

Mastication

Internal physical properties

Moisture-laden fruit fleshes are exceedingly difficult to classify. Some fleshes, of the Myristaceae (nutmeg family) and Lauraceae for example, are rich in lipids, while others are rich in carbohydrates (Leighton & Leighton, 1983). Textural measurement is again indicated but so far totally lacking. Moisture content would not appear to correlate with oral sensation. For example, *Ficus consociata* (Moraceae) and *Willughbeia coriacea* (Apocynaceae), with mean flesh moisture contents of 75–7%, do not release this moisture unless squeezed very hard. They taste slightly dry. In contrast, *Mezzettia leptopoda* (Annonaceae) and *Calophyllum pulcherrimum* (Guttiferae) with equivalent figures of 72–7% ooze fluid easily. The contrast would appear to be produced by the stiffness of the intercellular links in the latter two fruits with the result that individual cells are easily broken and release their contents.

In many fruits, the flesh is attached to the seed by fibres (e.g. *Mangifera indica*) which prevents flesh removal and encourages swallowing by a mammal. This is in sharp contrast to the ease with which most peels of protected fruits can be removed.

By far the most interesting of plant tissues in primate diets are seeds (which definition here includes any non-nutritive inner fruit layers when relevant) that house embryonic tissue that forms the potential next generation of plants. Whereas fruit tissue is designed to utilise available environmental agents (animals, wind, water) to disperse from the parent, the contained seed(s) must be highly insured against death. This insurance is often achieved mechanically. Contrasting solutions are

Fig. 3. Three seeds sectioned to show the extent of their woody coating (see the text). (a) *Mezzettia leptopoda* showing a woody plug to the left. Material for mechanical tests for this seed were from the flattened face. (b) *Neesia altissima* and (c) *Calamus* cf. *luridus*. Scale bar represents 10 mm.

shown in Fig. 3 where the thickness of the seed coat decreases left to right. However, the mechanical protection of each of these seeds by different means is considerable.

Mezzettia leptopoda seeds (Fig. 3a) are hemispherical and are contained inside a thickly fleshy fruit covered by a thin flexible skin. The seeds possess a 3–4 mm thick woody coat from which trabeculae extend into the endosperm. Through most of the thickness, the long thin cells, of 15–30 μm diameter, that form the coat are intermeshed in a complicated fashion. However, the outermost cells are oriented approximately perpendicular to the surface (i.e. radially). Its mechanical properties are summarised in Table 3, where it would appear that it is a tougher material than the shells of *Macadamia ternifolia* (Proteaceae). *Mezzettia* seeds appear to have defied Galdikas's (1982) attempts to open them with a machete despite the fact that she observed orang-utans (*Pongo pygmaeus*) opening them with their teeth and eating the oily endosperm. In order to ascertain whether the similar properties of the seed shells of distantly related *Mezzettia* and *Macadamia* were the rule, a seed with a thin shell was tested – *Calophyllum pulcherrimum* (Guttiferae). Values for all this unpublished data (Table 3) are intermediate to those for wood across and along the grain and similar to *Macadamia* and *Cocos*. Seed toughness is clearly variable but never equal to that of wood. Shells as thin as *Calophyllum* (Fig. 4) might be expected to have a low toughness because the options for structural complexity are limited by the size of the component fibres. However, because of the shape of most seeds, it is necessary to use C-ring tests (Jennings & Macmillan, 1986), which do not

Table 3. *Mechanical properties of seed shells compared to wood*

	Failure strength (MPa)	Elastic modulus (GPa)	Work to fracture kJm^{-2}
Mezzettia leptopoda (Annonaceae)	67	7.0	1.9
Ricinodendron rautanenii[a] (Euphorbiaceae)	58	4.3	—
Calophyllum pulcherrimum (Guttiferae)	17–22	1–1.8	0.11
Cocos nucifera[b] (Palmae)	—	2.9–4.9	1.7–1.9
Macadamia ternifolia[c] (Proteaceae)	25–80	2–6	0.1–1[d]
Wood[e]	5–100	0.5–10	0.1–10

Work to fracture estimates are referred to one side of the crack only.
[a]P. W. Lucas & C. R. Peters (unpublished data).
[b]J. F. V. Vincent & G. Jeronimidis (unpublished data).
[c]Jennings & Macmillan (1986).
[d]Estimated by Jennings & Macmillan (1986).
[e]Jeronimidis (1980).

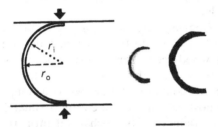

Fig. 4. Most seeds must be tested as C-rings in compression, both intact and notched (tests described by Jennings & Macmillan, 1986). Strictly, the sum of the outer (r_0) and inner (r_i) radii of curvature should exceed their difference by a factor of 20. This is so for *Calophyllum pulcherrimum* (left) but not for the post-test specimen of *Ricinodenron rautanenii* (right). The latter was not notched and failed in the centre; it required flats on the ends of the rings to prevent slippage. Despite appearances (drying flesh), the *Calophyllum* seed coat is of fairly constant thickness and does not possess a woody plug (Stevens, 1980) as does *Mezzettia* (Fig. 3). Scale bar represents 10 mm.

allow a direct work of fracture measurement, complicating comparisons (Fig. 4).

Some seeds, such as *Neesia altissima* (Fig. 3b) are protected by a thin non-woody coat which is flexible and rubbery and fails in tension easily.

The coat is about an order of magnitude lower in modulus than the *Mezzettia* shells (P. W. Lucas, unpublished data). They were not eaten by *Macaca fascicularis* probably because they are protected mechanically by stiff whisker-like projections from the inner surface of the dehisced fruit wall. These whiskers are several millimetres long and average 60 μm in diameter, with a radius of tip curvature (sharpness) of the order of 1 μm (Fig. 1). They penetrate the skin with ease and break off causing acute irritation.

The seeds of *Calamus* cf. *luridus* (Palmae) possess a very thin insignificant woody covering that sends trabeculae through the endosperm (Fig. 3c). Seeds of related fruit, *Plectocomia griffithii*, were successfully fractured by captive *M. fascicularis* but wild animals were not seen to do this. Rattan seeds are apparently rigid and tough. This is due to the endosperm, which appears similar to that of vegetable ivory. It was impossible to test the mechanical properties of this *Calamus* fruit due to its trabeculation and size. However, for many other seeds it is only the endosperm, possibly fairly tough (Vincent, 1990), that could deter predation on mechanical grounds because the woody coat is too thin to explain their apparent rigidity.

External physical properties

The distribution of seed sizes (measured again as maximum widths) for plant species in Bukit Timah is bimodal with two peaks, 0–1 mm and 6–8 mm. *Macaca fascicularis* fed on seeds across the size spectrum. The fate of seeds depended on whether the fruit was dry, when they were destroyed, or fleshy in which case the key to their fate was size (Corlett & Lucas, 1990).

A notable feature of the feeding behaviour of *M. fascicularis* was its tendency to chew very small mouthfuls at a time (the only exception being the consumption of *Xanthophyllum* leaves). The volume of food ingested may not be equivalent to the mouthful being chewed. Old World monkeys possess cheek pouches, which are muscular projections behind the molar teeth, in which unprotected or peeled fruit are stored. Judging from *M. fascicularis*, these items are apparently returned one by one to the oral cavity. We have linked this behaviour to the largely intra-oral removal of the fruit flesh followed by seed-spitting (Corlett & Lucas, 1990). The seeds of almost all fleshy fruits above 3–4 mm in maximum width were spat by *M. fascicularis*, whereas smaller ones are swallowed, which we believe forms the best evidence for a food particle size threshold for swallowing in any mammal.

Attempts to reproduce this behaviour in experiments on captive

Table 4. *Indentation hardness numbers (kg mm⁻²) of hard objects that enter the mouth*

Sclerotised cuticle (locust mandible)	18–36	Hillerton (1980)
Mezzetia leptopoda seed shell	Hv 21	P. W. Lucas (unpublished data)
Elaeis guineensis seed shell	Hv 21–24	P. W. Lucas (unpublished data)
Bone	Hv 50	P. W. Lucas (unpublished data)
Silica		
Quartz	Hk 710–790	Baker *et al.* (1959)
Opal phytoliths	Hk 590–610	Baker *et al.* (1959)

Hv, Vickers test; Hk, Knoop test.

Table 5. *Indentation hardness numbers (kg mm⁻²) of teeth*

	Enamel	Dentine	Author(s)
Man Hk	365–393	74	Waters (1980) Braden (1976)
Beaver [2] Hv		57	Osborn (1969)
Sheep Hk	270–382	33–74	Baker *et al.* (1959)
Orang-utan[a] Hk	362	49	P. W. Lucas (unpublished data)
Gibbon *(H. muelleri)*[a] Hk	282	59	P. W. Lucas (unpublished data)
Macaca nemestrina[a] Hk	386	60	P. W. Lucas (unpublished data)
Lungfish petrodentine Hv		318	Bemis (1984)
Trabecular dentine Hv		59	Bemis (1984)

Hk, Knoop test; Hv, Vickers test.
[a]Dried skulls.

animals succeeded in demonstrating a similar threshold but with the important difference that most seeds above the threshold were fractured and broken in the mouth rather than spat. It is possible that the wild populations from which these animals were caught had different habits but analysis of the erratic behaviour of individuals militates against this. For example, three male *M. fascicularis* monkeys, out of sight of each other, were each presented with 50 *Vitex pinnata* (Verbenaceae) fruits.

One spat all seeds intact, another apparently fractured all seeds (judged by the unmistakeable sounds of crispness) and swallowed them, whereas the last fractured all seeds and spat the fragments (Corlett & Lucas, 1990). It is unlikely that the latter behaviour could ever have evolved in a wild population. It is plausible that monkeys are socialised into their feeding behaviours, particularly with seeds that are tasteless but often highly toxic if fractured (Janzen, 1978; Bell, 1984). Seeds do not seem to give visual cues as to their toxicity. The spitting of seeds in the wild is known from *M. mulatta* (R. T. Corlett, unpublished data) and West African Old World monkeys (Gautier-Hion, 1980), all of which possess cheek pouches. Qualitative observations by Ridley (1930) on the feeding behaviour of another population of *M. fascicularis* in Singapore (now exterminated) support our results in the field.

The abrasiveness of foods is one element of the wear process. The only available data are on indentation hardness (Table 4). Comparing these values to those for dental tissues (Table 5) shows that only opal phytoliths found in some plants and quartz (grit) possess hardnesses greater than that of enamel. Nevertheless, it is known that keratins (Rose, Walker & Jacobs, 1981) and leaves (Walker, Hoeck & Perez, 1978) can wear teeth leaving them smooth.

Dental adaptation

Tooth shape and size

How can the above information be utilised to gain some insight into the dentitions of mammals? Assuming that the mechanical properties of mammalian teeth either do not vary or that variation is related only to durability (Janis & Fortelius, 1988), it is possible to predict some features of teeth. The action of the incisors can be difficult to analyse. Loading is variable and other teeth may also be used. The whole dentition of some fruit bats is disposed so that it can act in ingestion (Freeman, 1988). This is achieved by a very wide shallow dental arch and, presumably, a wide mouth-slit. The dentition of some primates, for example gibbons, resemble those of fruit bats (Freeman, 1988). Folivorous primates tend to have small incisors compared to their body weight (Hylander, 1975) possibly because leaves are flexible sheets that are easily folded into the mouth whatever their sizes. The canines, when they project beyond the other teeth, are generally used to apprehend prey or for fighting and have been analysed by beam analysis (Van Valkenburgh & Ruff, 1987). The upper size limit of the canines of males of anthropoid primate species appears to be governed by the size of the jaw and the angle to which it can be opened (Lucas *et al.*, 1986*b*) but many mammals may have canines smaller than predicted by this analysis. Female primates generally have relatively smaller canines than males but gibbons and the Pithecidae (three genera of New World monkeys) are exceptions. In one genus of pithecids, *Chiropotes*, the canines are used for opening up fruits (van Roosmalen, 1984).

To analyse the postcanine dentition, we start by returning to the statistical functions defined earlier, the selection (probability of fracture) and breakage (distribution of fragment size) functions. Because chewing is an interaction between oral and food surfaces, features of both foods and the mouth affect values of the selection and breakage functions. Lucas *et al.* (1986a) suggested that external food properties affect the selection function whereas internal properties, by definition not expressed at the food surface, primarily affect the breakage function. Many features of the mouth, particularly the actions of the tongue assisted by the cheeks, affect the selection function, whereas the breakage function is only a consequence of the pattern of loading imposed by tooth surfaces moved by muscles. The size of the working surface of the teeth is an important factor in determining the selection function, whereas tooth shape is more important for the breakage function (Lucas *et al.*, 1986a).

Lucas (1979) gave basic design features for molars comminuting solid foods. Assuming a need to minimise area of loading and wear, it is the rigidity (modulus) and fracture properties of foods that influence shape features. It is necessary to eliminate all food solids from the analysis that possess such a negligible strength that dental adaptation is unnecessary. If a food solid propagates cracks easily, then pointed structures (cusps) are indicated. If the food is rigid, then the area of loading will be very small (since a high stiffness to the teeth is an obvious prerequisite). Assuming that rigid foods are strong, then stress concentrations could be reduced by decreasing the sharpness (radius of curvature) of the cusp tip. This would do little to increase the area of loading and, judging from Auerbach's relationship for spherical indenters (Frank & Lawn, 1967), might decrease the stress at which the solid failed and reduce yielding that might otherwise inhibit crack growth. Since the crack would grow rapidly, the cusp tip is saved from wear because it would trail behind the crack perhaps analogously to the wedge test described by Vincent (1990). Designing a reciprocal concave-shaped basin (of radius of curvature greater than the cusp tip), allows the fragments to be collected in a position where multiple further fractures could result from a single movement of the cusp or basin.

If a food is reluctant to propagate cracks then an extension of one dimension of a pointed cusp to form a blade (ridge or crest), of dimensions greater than that of the food particle, is the optimum. Most foods that are reluctant to grow cracks are low modulus and therefore the blade would be supported by deformation of the food over its edges. There is therefore a necessity for the blade to be as sharp (low radius of curvature) as possible or the crack will be arrested. Dependent upon the conversion of strain energy in the food into fracture, it may be that the blade trails through the food slightly behind the crack tip which would reduce its wear. The result would be two fragments per parent food particle per blade. The reciprocal tooth surface must be a blade oriented in parallel otherwise the opposing surface will damage the blade.

The effective action of these double-bladed or pestle-and-mortar systems produces a non-zero value for the breakage function when used on the appropriate food and a zero value or lower than optimal in a mismatch. Unfortunately, no mammalian equivalent of Raupp's (1985) beetle experiment has been reported, the nearest being that of Berkovitz & Poole (1977). They found that hard brittle chow blunted the carnassial blades of ferrets whereas a diet of mice kept them sharp. However, they did not attempt to show that blunted carnassials slowed the processing of mice, which would be a key observation.

The radius of curvature of 'pestle' cusps will also affect the selection function because there will be a zone around the tip inside which particles possess a chance of being fractured. The size of this zone, for any given cusp shape, will depend on particle size (Lucas & Luke, 1984). The positions at which cusps are loaded, dependent on the modulus of the food, mouthful and food particle size, strongly influences features such as cusp height and shape (Preuschoft, 1989). In general, the number of pestles-and-mortars and blades in a dentition will vary the selection function and be determined by the temporal constraints on oral processing. Mastication may not always be a one-step process. The teeth on which different foods are chewed in man varies (Tornberg *et al.*, 1985). Hyaenids and canids use their carnassials for chewing the soft tissues of animals but other teeth for breaking bones.

Since fruits and seeds dominate primate diets, an attempt can be made to analyse their dentitions from this viewpoint. It is possible that the spatulate (bladed) incisor teeth distinctive of anthropoid primates could have evolved to cope with thick-peeled fruit (Lucas, 1989*a*). Most peels appear to lack fracture toughness, which would indicate pestle-and-mortar features rather than their double-bladed appearance. However, removal of fruit peels resembles sculpture rather than comminution in that crack propagation needs to be controlled so as not to risk dropping part of the fruit (which nevertheless macaques do all the time).

One type of analysis assigns a size to a tooth class on the basis of how much work it is required to perform (e.g. Hylander, 1975; Kay, 1975).

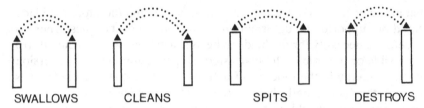

SWALLOWS   CLEANS   SPITS   DESTROYS

Fig. 5. Diagrammatic views of the working surface of the dentitions that are predicted for seed swallowers, cleaners, spitters and destroyers (see the text). The canines are shown as constant-sized triangles. Adaptation to these modes of seed processing is predicted to be achieved by altering the sizes of the postcanines (solid line) and incisors (dotted line).

Assuming that the length of the blade of incisors and the area of the working surface of the postcanines measures size, then Lucas (1989a) has suggested the following patterns of tooth sizes with diets (Fig. 5) based on size thresholds:

1. Some primates peel fruits and then swallow seed(s) and flesh. These primates do the minimum and it would be expected that they have the smallest incisors and postcanines.
2. Some primates peel fruits, swallow the seed(s) and flesh of most but clean the flesh from large seeds with the incisors. These primates have larger incisors but equally small postcanines.
3. Some primates peel fruits but clean the flesh from large seeds with the incisors and from smaller seeds with the post-canines. The seeds are then spat. These primates have large incisors and postcanines. Blunt cusps on the postcanine teeth would maximise the number of turgid cells in a stiff flesh (such as *Calophyllum* or *Mezzettia*) that could be broken per chew. If the flesh yields too much (e.g. most *Ficus*) then the importance of a seal between cusp and fossa (pestle-and-mortar) with restricted exits becomes important to try to break cells. This is the compression chamber of Maier (1977). Even though such primates do not break many seeds, the constant contact of seed shells with the teeth could be important for wear studies. The advantages of spitting are discussed by Corlett & Lucas (1990).
4. Some primates peel fruits and then destroy the seeds still with flesh attached with the postcanines. These primates have incisors as (1) but postcanines as (3).

This classification only considers fruit-eating. Generally, leaf-eating primates are the only species that consistently fracture seeds in the wild, which is connected to their microbial stomachs (Waterman, 1984). Some of these leaf-eaters may prefer seeds to leaves when the former are available (Davies & Baillie, 1988). Lucas *et al.* (1986*a*) suggested that fruit tissue would form a more cohesive bolus than other plant foods such as leaves. The yielding nature of the intercellular links in some fruit fleshes coupled with the admixture of latex in some species might support this. Further points are discussed by Lucas (1989*a*).

Usually the function of the postcanine teeth depends on their being covered by the cheeks, so that the width of the mouth-slit is, at maximum, probably related to the width of the anterior teeth. This may restrict the size of the canine teeth in male primates, which is exceedingly large for modern mammals (Lucas *et al.*, 1986*b*).

Properties of teeth

The physical properties of human teeth have been reviewed by Waters (1980). There are few comparable data for other mammals except for indentation hardness (Table 5). Assuming Knoop and Vickers tests to give similar readings (the former may be higher), mammalian enamel would appear to possess a fairly consistent hardness which is often slightly greater at the tooth surface. However, there are no published data on the hardness of pigmented enamels which contain goethite (Akersten, Lowenstam & Walker, 1984), which give the radula of molluscs a Vickers hardness number of about 600 (Runham *et al.*, 1969). Dentine hardness varies with distance from the enamel-dentine junction and pulp cavity. The dentine within 100 μm of the enamel is distinctly softer in man (Renson & Braden, 1971), the incisors of the beaver, *Castor fiber* (Osborn, 1969), and a gibbon, *Hylobates muelleri* (P. W. Lucas, unpublished data). In some lungfish, dentine can be mineralised as heavily as enamel (Bemis, 1984). Of importance for tooth shape, Osborn (1969) showed that the distribution of dentine hardnesses differed in upper and lower incisors of beavers and that this was correlated with the wearing of distinct steps in the upper incisors of functional importance. According to Osborn, these steps are not found in all rodents. They are also found variably in primates (Teaford, 1988).

Much is known about enamel structure and this varies considerably between mammalian groups. Pre-fracture toughness (ability to store strain energy) may be enhanced by the microporosity of the enamel which renders water flowing through the pores more viscous by an electrical effect (Fox, 1980). Fox speculates that hydrophilic organic material attracts water back on unloading. Nothing is known of how enamels may vary in this respect. The anisotropy of the work of fracture of enamel (Rasmussen *et al.*, 1976) is produced by a crystallite-free boundary between rod-shaped aggregations of crystals called prisms, which are often about 5 μm in diameter (Shellis, 1984). The work of fracture is much greater across prisms than between them. The variety of prism directions in different enamels must affect resistance to fracture but most studies have inferred this and not tested it in any systematic way (reviewed by Janis & Fortelius, 1988).

Wear

Wear deserves special consideration because, even though difficult to quantify and impossible to predict in any specific way, it is currently the prime focus for dental-dietary studies (Teaford, 1988). Early studies concentrated on wear facets for predicting jaw movements (Mills, 1955). It is still unresolved as to whether this is possible. Later, many smoother wear facets were interpreted as being produced by deliberate jaw movements, made without food in the mouth, for the purpose of keeping teeth sharp (Every, 1970). Though there are exceptions (Osborn, 1969; Walker, 1984), this is generally unlikely (Teaford & Walker, 1983). Some carnivorous vertebrates such as sharks, have enamelloid-capped teeth that are so sharp as to fracture human skin with ease (dentition analysed by Preuschoft, Reif & Muller, 1974). Piranhas also have equally sharp teeth though their diets often include plant matter (Nico & Taphorn, 1988).

Land mammals never have teeth so sharp. This could be a consequence of their not being continually replaced as is the rule in other vertebrate dentitions but even an unworn carnassial in the domestic cat possesses a radius of curvature of the blade in the tens of micrometres (P. W. Lucas, unpublished data). It is tempting to relate this to the possession of prismatic enamel, rare in non-mammalian vertebrates, since sharpness must be limited by prism diameters. The cost in reducing crack propagation within enamel by forming prisms may be a higher abrasion rate in relation to mineral density (cf. mollusc shells; Currey, 1980) and a consequent loss of ability to maintain sharpness.

If tooth wear is mostly of an abrasive type (Bowden & Tabor, 1973), then relative indentation hardnesses might indicate the rate of tissue loss. Table 4 shows that only siliceous materials, found particularly in grass roots (MacNaughton *et al.*, 1985), the storage organs of which are eaten extensively by *Papio* baboons, are likely to cause heavy wear. Seed shells have a hardness about the same as that of the dentine adjacent to the

Table 6. *pH of fruit flesh*

Calophyllum pulcherrimum[a] (Guttiferae)	3.0–3.5
Garcinia cf. *forbesii*[a] (Guttiferae)	2.5–3.0
G. parvifolia[a] (Guttiferae)	3.5–3.0
Nephelium lappaceum[a] (Sapindaceae)	4.6
Tinomiscium petiolare Menispermaceae)	5.0–5.5
Fibrauria tinctoria (Menispermaceae)	5.5–6.0
Mezzettia leptopoda (Annonaceae)	4.0–4.5
Tetrastigma lawsoni[a] (Vitaceae)	3.0
Pellacalyx saccardianus[b] (Rhizophoraceae)	5.0
Gnetum microcarpum (Gnetaceae)	5.0

Measured with Half-unit pH paper.
[a]Genera important in primate diets in region.
[b]Fruit avoided by *Macaca fascicularis*.

tooth pulp and are unlikely to scratch enamel (as shown in the experiments of Peters, 1982) but may polish it (Peters, 1987).

Being composed of hydroxyapatite, teeth are vulnerable to acid erosion. Acids of a pH less than 5.5–6.0 can dissolve enamel or dentine at a rate dependent on mineral content. Elsbury (1952) found the ability of acid to reduce enamel volume to be greatest at a pH below 3.8, but the microhardness of human enamel diminishes significantly after exposure to a pH of 4.0 and may decline at a faster rate than loss of calcium (Davidson, Hoekstra & Arends, 1974). Erosion by food acid could wear dental tissues substantially and interfere with scratch analysis (Teaford, 1988) by etching off marks left by abrasion.

Many fruits are apparently protected by acidity when unripe (Ridley, 1894) and some remain acid, disguised by sugars, when fully ripe. Fruits are the likeliest source of acid and two of the commonest are malic and citric acid (Wills *et al.*, 1981); the latter is particularly damaging to enamel (Elsbury, 1952). Sourd & Gautier-Hion (1986) mention two ripe fruit species eaten by *Cercopithecus cephus* that were very acidic as assessed by titration but there appear to have been no previous field reports of the pH of food items. Therefore unpublished acidities of some ripe fruits in Bukit Timah are given in Table 6, which show that considerable erosion of the teeth by some fruits is possible.

Change in pH during the development of the rambutan, *Nephelium lappaceum* (Sapindaceae) fruit has been studied by Poon (1974). This fruit is cultivated but is also common as a wild forest fruit in forests; it is one of the key food genera for Southeast Asian macaques (Caldecott, 1986). The fruit takes about 14 weeks to ripen. During weeks 6–14, the pH rises from 3.0 to 4.6 as acids are converted to sugars. The ripe fruits of many wild species taste noticeably sourer than *N. lappaceum*, which may indicate that conversion of acids to sugars is less than in cultivated forms. A pH of 1.5–2.0 was found in full-size *Garcinia* cf. *forbesii* (Guttiferae) fruit, which were green rather than their ripe orange-brown. Interestingly, the titratable acidity (mostly citric acid) of the cultivated mangosteen, *Garcinia mangostana*, peaks at about 90% of final fruit size (Tan, 1964) though pH is unknown. Apparently ripe *Garcinia parvifolia* (Guttiferae) and *Tetrastigma lawsoni* (Vitaceae) fruit were eaten by *M. fascicularis* at a pH of 2.5–3.0, the lowest pH values that we have ascertained in its diet. Both plant genera are important in primate diets elsewhere in Southeast Asia (Caldecott, 1986; Leighton & Leighton, 1983).

Dental adaptations to wear can be summarised as follows:

1. Abrasive wear by foods in chewing can be decreased by increasing tooth size. This is because most available loading sites are not utilised in a chew and food is constantly recirculated (see above). Therefore increasing the number of sites decreases the wear at any given site per unit of time.

2. Abrasive wear of teeth by each other (called attrition in the literature) is unaffected by change in tooth size; tooth parts always contact at the same sites. However, the life of a tooth can be extended to counteract such wear by thickening the tissue.

3. Unlike (1) and (2), erosion can affect the whole tooth crown. It is the only method by which the, otherwise stable, maximum bucco-lingual width of a tooth could be reduced and features such as perikymata be lost. However, change in salivary pH would seem the obvious adaptation to combat food acidity. Puech (1984) has published a micrograph of an orang-utan molar which may be eroded. Eroded enamel is liable to be abraded easier (Teaford, 1988). Lucas (1989*b*) has suggested that seed shells could then become very important in polishing enamel.

Conclusions

It may be because most human foods are often so weak that much of dental science has ignored food texture; indeed the emphasis on occlusion (tooth contact) in dentistry is appropriate if the forces when teeth contact during chewing and swallowing are much greater than those when fracturing food. Teeth are often the hardest things in the mouth and reflexes that switch off jaw closing muscles and prevent uppers damaging lowers are activated only after tooth contact (Olthoff, 1986). In palaeontology, little other than studies of occlusion or interpretations of wear by analogy with living forms can be achieved in morphological studies. In ecology, the influence of texture on food acquisition and processing has been unfairly overlooked. We hope that textural study of the natural foods of living mammals in conjunction with oral physiology will make significant advances in understanding dentitions.

Acknowledgments

We thank E. J. Ang, B. Pereira & J. Goh (Orthopaedic Surgery), D. Lane and D. H. Murphy (Zoology), Mr Samsuri, T. Lowrey, T. C. Whitmore and H. Tan (Botany), Mrs C. M. Yang and Miss H. K. Lua (Zoological Reference Collection), B. M. A. Tajuddin, H. L. Chan and Gopal (Anatomy) all of the National University of Singapore; M. C. Ng and S. M. Tan of the Marine Fisheries Research Department, Southeast Asian Fisheries Development Centre, Singapore and S. Richardson and the staff of the Metallurgy Laboratory, Nanyang Technological Institute, Singapore.

References

Akersten, W. A., Lowenstam, H. A. & Walker, A. C. (1984). 'Pigmentation' of soricine teeth: composition, ultrastructure, and function. Abstract. *American Society of Mammalogists, 64th Annual Meeting*, no. 153, 40.

Baer, H. (1983). Allergic contact dermatitis from plants. In *Handbook of Natural Toxins*, vol. 1 *Plant and Fungal Toxins*, ed. R. F. Keeler & A. T. Tu, pp. 421–42. Marcel Dekker, New York.

Baker, G., Jones, L. H. P. & Wardrop, I. D. (1959). Cause of wear in sheep's teeth. *Nature (London)*, **184**, 1583–4.

Bell, E. A. (1984). Toxic compounds in seeds. In *Seed Physiology*, ed. D. R. Murray, vol. 1, pp. 245–64. Academic Press, Sydney.

Bemis, W. E. (1984). Morphology and growth of lepidosirenid lungfish tooth plates (Pisces: Dipnoi). *Journal of Morphology*, **179**, 73–93.

Berkovitz, B. K. B. & Poole, D. F. G. (1977). Attrition of the teeth in ferrets. *Journal of Zoology*, **183**, 411–18.

Bowden, F. P. & Tabor, D. (1973). *Friction: An Introduction to Tribology*. Doubleday, New York.

Braden, M. (1976). Biophysics of the tooth. In *Frontiers of Oral Physiology*, ed. Y. Kawamura, vol. 2, pp. 1–37. Karger, Basel.

Byrd, K.E., Milberg, D.J. & Luschei, E.S. (1978). Human and macaque mastication: a quantitative study. *Journal of Dental Research*, **57**, 834–43.

Caldecott, J.O. (1986). An ecological and behavioural study of the pig-tailed macaque. *Contributions to Primatology*, **4**, 1–335.

Coley, P.D. (1983). Herbivory and defensive characteristics of tree species in a lowland tropical forest. *Ecological Monographs*, **53**, 209–35.

Coley, P.D. (1987). Interspecific variation in plant anti-herbivore properties: the role of habitat quality and rate of disturbance. *New Phytologist*, **106** (supplement), 251–63.

Corlett, R.T. (1988). Bukit Timah: history and significance of a small rain-forest reserve. *Environmental Conservation*, **15**, 37–44.

Corlett, R.T. (1990). Flora and reproductive phenology of a Singapore rain forest. *Journal of Tropical Ecology*, **6**, 55–63.

Corlett, R.T. & Lucas, P.W. (1990). Alternative seed-handling strategies by primates: seed-spitting by *Macaca fascicularis*. *Oecologia*, **82**, 166–71.

Currey, J.D. (1980). Mechanical properties of mollusc shell. In *The Mechanical Properties of Biological Materials*, Symposia of the Society for Experimental Biology, no. 34, ed. J.F.V. Vincent & J.D. Currey, pp. 75–97. Cambridge University Press, Cambridge.

Davidson, C.L., Hoekstra, I.S. & Arends, J. (1974). Microhardness of sound, decalcified and etched tooth enamel related to the calcium content. *Caries Research*, **8**, 135–44.

Davies, A.G. & Baillie, I.C. (1988). Soil-eating by red-leaf monkeys (*Presbytis rubicunda*) in Sabah, Northern Borneo. *Biotropica*, **20**, 252–8.

Elsbury, W.B. (1952). Hydrogen-ion concentration and acid erosion of the teeth. *British Dental Journal*, **93**, 177–9.

Every, R.F. (1970). Sharpness of teeth in man and other primates. *Postilla*, **143**, 1–20.

Fox, P.G. (1980). The toughness of tooth enamel, a natural fibrous composite. *Journal of Materials Science*, **15**, 3113–21.

Frank, F.C. & Lawn, B.R. (1967). On the theory of Hertzian fracture. *Proceedings of the Royal Society of London*, **299**, 291–306.

Freeman, P.W. (1988). Frugivorous and animalivorous bats (Microchiroptera): dental and cranial adaptations. *Biological Journal of the Linnaean Society*, **33**, 249–72.

Galdikas, B. (1982). Orang-utans as seed dispersers at Tanjung Puting, Central Kalimantan: implications for conservation. In *The Orang Utan. Its Biology and Conservation*, ed. L.E.M. de Boer, pp. 285–98. W. Junk, The Hague.

Gautier-Hion, A. (1980). Seasonal variations of diet related to species and sex in a community of *Cercopithecus* monkeys. *Journal of Animal Ecology*, **49**, 237–69.

Harborne, J. (1982). *Introduction to Ecological Biochemistry*, 2nd edn. Academic, London.

Heath, M. R. & Lucas, P. W. (1988). Oral perception of texture. In *Food Structure – Its Creation and Evaluation*, ed. J. M. V. Blanchard & J. R. Mitchell, pp. 465–81. Butterworths, London.

Herring, S. W. (1976). The dynamics of mastication in pigs. *Archives of Oral Biology*, **21**, 473–80.

Hiiemae, K. M. & Crompton, A. W. (1985). Mastication, food transport and swallowing. In *Functional Vertebrate Morphology*, ed. M. Hildebrand, D. M. Bramble, K. F. Liem & D. B. Wake, pp. 262–90. Belknap, Harvard.

Hillerton, J. E. (1980). The hardness of locust incisors. In *The Mechanical Properties of Biological Materials*, Symposia of the Society for Experimental Biology, no. 34, ed. J. F. V. Vincent & J. D. Currey, pp. 483–4. Cambridge University Press, Cambridge.

Hylander, W. L. (1975). Incisor size and diet in anthropoids with special reference to Cercopithecoidea. *Science*, **189**, 1095–7.

Hylander, W. L., Johnson, K. R. & Crompton, A. W. (1987). Loading patterns and jaw movements during mastication: a bone-strain, electromyographic and cineradiographic analysis. *American Journal of Physical Anthropology*, **72**, 287–314.

Janis, C. M. & Fortelius, M. (1988). On the means whereby mammals achieve increased functional durability of their dentitions, with special reference to limiting factors. *Biological Reviews*, **63**, 197–230.

Janson, C. H. (1983). Adaptation of fruit morphology to dispersal agents in a neotropical rain forest. *Science*, **219**, 187–9.

Janzen, D. H. (1978). The ecology and evolutionary biology of seed chemistry as relates to seed predation. In *Biochemical Aspects of Plant and Animal Evolution*, ed. J. B. Harborne, pp. 163–206. Academic, London.

Janzen, D. H. (1985). Plant defences against animals in the Amazonian rainforest. In *Key Environments – Amazonia*, ed. G. T. Prance, pp. 207–17. Pergamon, Oxford.

Jennings, J. S. & Macmillan, N. H. (1986). A tough nut to crack. *Journal of Materials Science*, **21**, 1517–24.

Jeronimidis, G. (1980). Wood, one of nature's challenging composites. In *The Mechanical Properties of Biological Materials*, Symposia of the Society for Experimental Biology, no. 34, ed. J. F. V. Vincent & J. D. Currey, pp. 169–82. Cambridge University Press, Cambridge.

Kay, R. F. (1975). Functional adaptations of primate molar teeth. *American Journal of Physical Anthropology*, **43**, 195–216.

Lanyon, J. M. & Sanson, G. D. (1986). Koala (*Phascolarctos cinereus*) dentition and nutrition. II. Implications of tooth wear in nutrition. *Journal of Zoology*, ser. A, **209**, 169–81.

Leighton, M. & Leighton, D. R. (1983). Vertebrate responses to fruiting seasonality within a Bornean rain forest. In *Tropical Rain Forest:*

Ecology and Management, ed. S.L. Sutton, T.C. Whitmore & A.C. Chadwick, pp. 181–96. Blackwell, Oxford.

Lucas, P.W. (1979). The dental–dietary adaptations of mammals. *Neues Jahrbuch für Geologie und Paläontogie Monatschafte*, **8**, 486–512.

Lucas, P.W. (1989a). A new theory relating seed processing by primates to their relative tooth sizes. In *The Growing Scope of Human Biology, Proceedings of the Australasian Society for Human Biology 2*, ed. L.H. Schmitt, L. Freedman & N.W. Bruce, pp. 37–49. Centre for Human Biology, University of Western Australia, Perth, WA.

Lucas, P.W. (1989b). Significance of *Mezzettia leptopoda* fruits eaten by orang-utans for dental microwear analysis. *Folia Primatologica*, **52**, 185–90.

Lucas, P.W. & Luke, D.A. (1983a). Methods of analysing the breakdown of food in human mastication. *Archives of Oral Biology*, **28**, 813–19.

Lucas, P.W. & Luke, D.A. (1983b). Computer simulation of the breakdown of carrot particles during human mastication. *Archives of Oral Biology*, **28**, 821–6.

Lucas, P.W. & Luke, D.A. (1984). Optimum mouthful for food comminution in human mastication. *Archives of Oral Biology*, **29**, 205–10.

Lucas, P.W., Corlett, R.T. & Luke, D.A. (1986a). Postcanine tooth size and diet in anthropoid primates. *Zeitschrift für Morphologie und Anthropologie*, **76**, 253–76.

Lucas, P.W., Corlett, R.T. & Luke, D.A. (1986b). Sexual dimorphism of tooth size in anthropoids. In *Sexual Dimorphism in Living and Fossil Primates*, ed. M. Pickford & B. Chiarelli, pp. 23–39. Il Sedicesimo, Florence.

Lumsden, A.G.S. & Osborn, J.W. (1977). The evolution of chewing: a dentist's view of palaeontology. *Journal of Dentistry*, **5**, 269–87.

Macarthur, C. & Sanson, G.D. (1988). Tooth wear in eastern grey kangaroos (*Macropus giganteus*) and western grey kangaroos (*Macropus fuliginosus*), and its potential influence on diet selection, digestion and population parameters. *Journal of Zoology*, **215**, 491–504.

MacNaughton, S.J., Tarrants, J.L., MacNaughton, M.M. & Davis, R.H. (1985). Silica as a defense against herbivory and a growth promotor in African grasses. *Ecology*, **66**, 528–35.

Maier, W. (1977). Die evolution der bilophodonten molaren der Cercopithecoiden. *Zeitschrift für Morphologie und Anthropologie*, **68**, 26–56.

Mills, J.R.E. (1955). Ideal dental occlusion in the primates. *Dental Practitioner*, **6**, 47–63.

Milton, K. (1980). *The Foraging Strategy of Howler Monkeys*. Columbia University Press, New York.

Murphy, D.H. (1973). Animals in the forest ecosystem. In *Animal Life and Nature in Singapore*, ed. S.H. Chuang, pp. 53–73. Singapore University Press, Singapore.

Ng, F.S.P. (1988). Forest tree biology. In *Key Environments – Malaysia*, ed. Earl of Cranbrook, pp. 102–25. Pergamon, Oxford.

Nico, L.G. & Taphorn, D.C. (1988). Food habits of piranhas in the low Llanos of Venezuela. *Biotropica*, **20**, 311–21.

Olthoff, L.W. (1986). Comminution and neuromuscular mechanisms in human mastication. Doctoral dissertation, University of Utrecht.

Osborn, J.W. (1969). Dentine hardness and incisor wear in the beaver (*Castor fiber*). *Acta Anatomica*, **72**, 123–32.

Peck, A.L. (transl.) (1937). *Aristotle: [Parts of Animals]*. Heinemann, London.

Peters, C.R. (1982). Electron-optical microscopic study of incipient dental microdamage from experimental seed and bone crushing. *American Journal of Physical Anthropology*, **57**, 283–301.

Peters, C.R. (1987). Nut-like oil seeds: food for monkeys, chimpanzees, humans and probably ape-men. *American Journal of Physical Anthropology*, **73**, 333–63.

Poon, T.F. (1974). Physiological studies on fruits of *Nephelium lappaceum* L. B.Sc. dissertation, Department of Botany, National University of Singapore.

Preuschoft, H. (1989). Biomechanical approach to the evolution of the facial skeleton of hominoid primates. *Fortschritte der Zoologie*, in press.

Preuschoft, H., Reif, W.E. & Müller, W.H. (1974). Funktionsanpassungen in Form und Struktur an Haifischzahnen. *Zeitschrift für Anatomie, Entwicklunggeschichte*, **143**, 315–44.

Puech, P.-F. (1984). Acidic-food choice in *Homo habilis* at Olduvai. *Current Anthropology*, **25**, 349–50.

Rasmussen, S.T., Patchin, R.E., Scott, D.B. & Heuer, A.H. (1976). Fracture properties of human enamel and dentine. *Journal of Dental Research*, **55**, 154–64.

Raupp, M.J. (1985). Effects of leaf toughness on mandibular wear of the leaf beetle, *Plagiodera versicolora*. *Ecological Entomology*, **10**, 73–9.

Renson, C.E. & Braden, M. (1971). The experimental deformation of human dentine by indenters. *Archives of Oral Biology*, **16**, 563–72.

Ridley, H.N. (1894). On the dispersal of seeds by mammals. *Journal of the Straits British Royal Asiatic Society*, **25**, 11–32.

Ridley, H.N. (1930). *The Dispersal of Plants Around the World*. Reeve, Kent.

Rose, K.D., Walker, A.C. & Jacobs, L.L. (1981). Function of the mandibular tooth comb in living and extinct mammals. *Nature (London)*, **289**, 583–5.

Runham, N.W., Thornton, P.R., Shaw, D.A. & Wayte, R.C. (1969).

Mineralisation and hardness of the radular teeth of the limpet *Patella vulgata*. *Zeitschrift für Zellforschung und Mikroscopische Anatomie Abteilung Histochemie*, **99**, 608–26.

Shaw, D. M. (1917). Form and function of teeth: a theory of 'maximum shear'. *Journal of Anatomy*, **52**, 97–106.

Shellis, R. P. (1984). Inter-relationships between growth and structure of enamel. In *Tooth Enamel IV*, ed. R. W. Fearnhead & S. Suga, pp. 467–71. Elsevier/North Holland, Amsterdam.

Sherman, P. (1969). A texture profile of foodstuffs based on well-defined rheological properties. *Journal of Food Science*, **34**, 458–62.

Simpson, G. G. (1936). Studies of the earliest mammalian dentitions. *Dental Cosmos*, **78**, 940–53.

Sourd & Gautier-Hion, A. (1986). Fruit selection by a forest guenon. *Journal of Animal Ecology*, **55**, 235–44.

Stevens, P. F. (1980). A revision of the Old World species of *Calophyllum* (Guttiferae). *Journal of the Arnold Arboretum*, **61**, 117–424.

Szczesniak, A. S. (1963). Classification of textural characteristics. *Journal of Food Science*, **28**, 385–9.

Tan, S. B. (1964). A preliminary investigation on certain aspects of fruit physiology of *Garcinia mangostana* L. B.Sc. dissertation, Department of Botany, National University of Singapore.

Teaford, M. (1988). A review of dental microwear and diet in modern mammals. *Scanning Microscopy*, **2**, 1149–66.

Teaford, M. & Walker, A. C. (1983). Dental microwear in adult and still-born guinea-pigs (*Cavia porcellus*). *Archives of Oral Biology*, **28**, 1077–81.

Terborgh, J. (1983). *Five New World Primates*. Princeton University Press, Princeton, NJ.

Tornberg, E., Fjelkner-Modig, S., Ruderus, H., Glantz, P.-O., Randow, K. & Stafford, G. D. (1985). Clinically recorded masticatory patterns as related to the sensory evaluation of meat and meat products. *Journal of Food Science*, **50**, 1059–66.

Van Roosmalen, M. G. M. (1980). Habitat preferences, diet, feeding strategy and social organisation of the black spider monkey (*Ateles paniscus paniscus* Linnaeus 1758) in Surinam. Doctoral dissertation, Rijksuniversiteit voor Natuurbeheer, Leersum.

Van Roosmalen, M. G. M. (1984). Subcategorizing foods in primates. In *Food Acquisition and Processing in Primates*, ed. D. J. Chivers, B. A. Wood & A. Bilsborough, pp. 167–75. Plenum Press, New York.

Van Valkenburgh, B. & Ruff, C. (1987). Canine tooth strength and killing behaviour in large carnivores. *Journal of Zoology*, **212**, 379–87.

Vincent, J. F. V. (1990). Fracture in plants. *Advances in Botanical Research*, **17**, 235–87.

Voon, F. C. T., Lucas, P. W., Chew, K. L. & Luke, D. A. (1986). A simulation approach to understanding the masticatory process. *Journal of Theoretical Biology*, **119**, 251–62.

Walker, A. C. (1984). Mechanisms of honing in the male baboon canine. *American Journal of Physical Anthropology*, **65**, 47–60.

Walker, A. C., Hoeck, H. & Perez, L. (1978). Microwear of mammalian teeth as an indicator of diet. *Science*, **1201**, 908–10.

Waterman, P. G. (1984). Food acquisition and processing as a function of plant chemistry. In *Food Acquisition and Processing in Primates*, ed. D. J. Chivers, B. A. Wood & A. Bilsborough, pp. 177–211. Plenum Press, New York.

Waters, N. E. (1980). Some mechanical and physical properties of teeth. In *The Mechanical Properties of Biological Materials*, Symposia of the Society for Experimental Biology, no. 34, ed. J. F. V. Vincent & J. D. Currey, pp. 99–135. Cambridge University Press, Cambridge.

Whitmore, T. C. (1984). *Tropical Rain Forests of the Far East*, 2nd edn. Oxford University Press, Oxford.

Whitten, A. J. (1982). Diet and feeding behaviour of Kloss gibbons on Siberut Island, Indonesia. *Folia Primatologica*, **37**, 177–208.

Wictorin, L., Hedegard, B. & Lundberg, M. (1968). Masticatory function – a cineradiographic study. III. Position of the bolus in individuals with full complement of natural teeth. *Acta Odontologica Scandinavica*, **26**, 213–22.

Wills, R. B. H., Lee, T. H., Graham, D., McGlasson, W. B. & Hall, E. G. (1981). *Postharvest*. New South Wales University Press, Sydney.

Yurkstas, A. A. & Curby, W. A. (1953). Force analysis of prosthetic appliances during function. *Journal of Prosthetic Dentistry*, **3**, 82–7.

E. OTTEN

The control of movements and forces during chewing

When one observes a mammal feeding, its chewing movements give the impression of being well planned, automatic and highly functional. Direct observation, however, does not uncover the immense problems in controlling forces and movements of the jaw, since most of the control parameters (the recruitment patterns of as many as 25 muscle units) remain invisible to the naked eye. A combination of measuring techniques and computer simulation should reach greater depth, although both can be developed and used only under guidance of well-formulated concepts of motor control.

Currently, the field of motor control is largely concentrated on the orchestration of signals, from the nervous system to the periphery and vice versa, which appears to be so complex that the mechanical properties of the muscles and bone–connective-tissue systems involved are not taken into account. A quite different approach is to describe a motor control system as a closed pathway of signals, with branches to the external world. In such a description, the physical expression of the signals changes through the circuit from electrical to mechanical and vice versa, so that the properties of the mechanical section of the pathway play their role in the explanation of motor control. The mechanical properties depend on the architecture of muscles and the bone–connective-tissue system down to the level of filaments. Since this architecture varies according to the space available and the functional demands imposed, it is pointless to use standard values for parameters which describe muscle architecture when calculating movements and forces from recruitment patterns of muscles. Either one measures length–force relations and length–velocity relations of muscles and trajectories of possible movements of the bone–connective-tissue system, or one designs testable models that calculate these properties from the architecture. The advantage of the latter method is that properties that are difficult or unethical to measure can be calculated from, for instance, nuclear magnetic resonance or tomography series.

Sarcomeres

Active movement is possible in the animal kingdom only because of the well-studied contractile mechanism of muscles (Pollack & Sugi, 1984). We formulated a mathematical model of the contraction process in sarcomeres of which the lengths of the actin and myosin filaments in a sarcomere, and the length of the cross-bridge-free zone of the myosin filament, were the constituent parameters (Otten, 1987a). The length–force diagram predicted by this model was well in agreement with the measured one, indicating realism of the model. It is useful to have a model of sarcomeres, since sarcomere architecture forms the basis of muscle properties and the lengths of the filaments of sarcomeres vary among species and even within one muscle, depending on the task of motor units (Akster et al., 1984). Moreover, it is simpler to measure sarcomere architecture than length–force diagrams of individual muscle fibres.

Skeletal muscles

Having unravelled the relation between form and function in sarcomeres, the way is open for establishing this relationship in whole muscles. If one makes an overview of all possible types of muscle architecture, there appears to be an enormous variation. Partly, this variation is due to the available space in the organism, but mostly the variation can be explained on functional grounds. This can be attempted by modelling skeletal muscles, starting with their architecture. The history of muscle models goes back to 1664, when the Danish scientist Stensen made simplified drawings of skeletal muscles, which indicate that he understood the principle of design of pinnate muscles as we understand it now (Scherz, 1963). As late as the twentieth century, new insights were gained from modelling (Alexander, 1968; Benninghoff & Rollhäuser, 1952; Gans, 1982; Gaspard, 1965; Pfuhl, 1937; Woittiez et al., 1984; Hatze, 1978; Heukelom et al., 1979; Gans & Bock, 1965; Otten, 1988). The latest models include volume retention during contraction, the principle of balance of work (the work produced by muscle fibres should be the same as the sum of the work taken up by tendinous sheets and tendon and the work produced externally by the muscle), stretch and curvature of tendinous sheets, curvature of the fibres, stretch of the tendon, unhomogeneity of the sarcomere and intramuscular pressure. These models have been corroborated by measurements of length–force diagrams of muscles and of intramuscular pressure (Otten, 1988).

Apart from giving insight into the functional meaning of muscle architecture, muscle models provide also the possibility of calculating functional characteristics (such as force development), where they are hard to measure, especially in human subjects.

The muscle–bone–ligament system

Mathematical descriptions of muscles have been incorporated into a model describing muscle–bone–ligament systems, such as the jaw system in rats (Otten, 1987*b*). If one is interested in movements, the dynamic properties of skeletal muscles come into play. These consist of well-studied force–velocity relations (Hill, 1938; Aubert, 1956) and the activation dynamics of muscles (Wallinga-DeJonge, 1980; Otten, 1987*b*). Muscle forces depend on activation, on muscle length and velocity of shortening. If these forces are calculated for all muscles in a system, and given the joint, connective tissue and bone characteristics which determine the kinematics and inertias, then in principle the movements of the system can be calculated. In reality, however, we are dealing with a very complex dynamic system with an irregular flow of mechanical energy coming in (from the muscles) and going out (to the external world).

A solution to this problem lies in its description as a problem of Hamiltonian mechanics for systems with a time-dependent Hamiltonian function (Arnold, 1978). (A numerical approximation of this solution is offered by Otten (1987*b*).) The Hamiltonian represents the total mechanical energy of the system, which determines the movements. Chewing is the mechanical result of complex recruitment patterns of jaw muscles. This result is not easily explained in a quantitative sense. One cannot simply look at the resulting set of muscle forces, add them all up and decide that the lower jaw will move in the direction of the sum vector.

Movements are usually not in the direction of muscle forces, because reaction forces occur and because muscle forces depend heavily on the velocity of contraction, which implies that the movement that is generated influences its own driving forces. Another problem is that muscles are not monotonic springs: their force is not always an increasing function of their length. This may result in unstable force equilibria. Figure 1 shows the difference between a simple sprung mechanism and a muscle suspended mechanism. When point M is moved away from the point where force equilibrium occurs (Fig. 1a), the mechanical energy (force integrated over displacement) increases by the area of the small black triangle. However, if the same is done in a muscle-suspended system (Fig. 1b), the mechanical energy decreases by the grey area due to the length–force relations of both muscles, implying that point M will

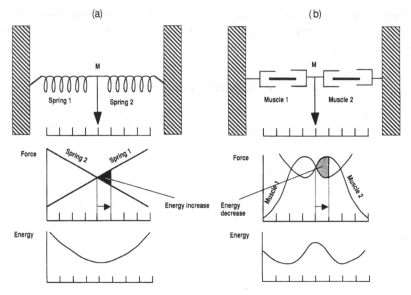

Fig. 1. (a) A sprung mechanism compared with (b) a muscle suspended mechanism. In the latter, an unstable force equilibrium occurs. The energy decrease indicated is equivalent to the decrease in the surface area under the two muscle length–force curves when point M is displaced away from the middle point where force equilibrium occurs.

move away from the point of force equilibrium. Therefore, in this situation the point of equilibrium is unstable and point M will move to one of the stable equilibria. This demonstrates that looking for force equilibria is not the way to find solutions for muscle-suspended mechanisms. One should look for the minima in mechanical energy. An advantage of this method is also that other forms of mechanical energy that are involved in the interaction with the external world or the food can easily be included in the objective function, the total mechanical energy.

Movements of a mechanism can be described in the space of the mechanical degrees of freedom of that mechanism. In the jaw system of the rat, six degrees of freedom can be distinguished: three degrees for the lower jaw (mouth opening and closing, jaw protrusion and retraction, and lateral movements of the jaw) and three degrees for the hyoid (the same as those for the lower jaw). Since the hyoid is very small, its rotations can be neglected, but its translations are important in understanding the motor control of the lower jaw. A six-dimensional space is very hard to draw, so the two most important degrees of freedom of the lower jaw are depicted in Fig. 2, together with the resulting mechanical

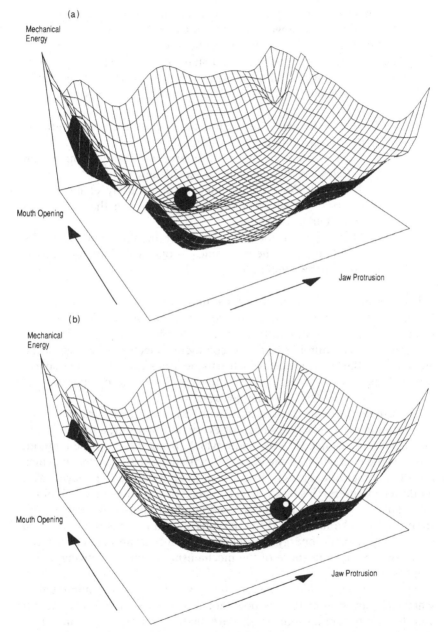

Fig. 2. (a, b) The total mechanical energy of the jaw system as a function of mouth opening and jaw protrusion for two different muscle recruitment patterns.

energy at two particular times. The energy landscapes in Fig. 2a and b differ slightly due to a change in the periodic muscle recruitment pattern. The black ball tries to follow the energetic minimum as closely as possible. The position of the ball defines a state of mouth opening and jaw protrusion; from a to b of Fig. 2, the lower jaw closes and protrudes partly. The mechanical energy consists of a number of components:

1. The strain energy of the muscles is calculated by integrating the muscle length force relations.
2. The potential (gravitational) energy of all bony elements and muscles can be calculated from the new positions.
3. The work done by the inertias of all bony elements and muscles can be calculated from changes in their momenta and angular momenta.
4. The energy needed to deform or break the food can be calculated from the mechanical properties of the food that is used in the simulation.

If one searches for the nearest local minimum after Δt in the energy function with newly updated recruitment rates of the muscles, using for instance a polytope optimisation algorithm (Nelder & Mead, 1965), and if one repeats this a number of times, complete trajectories emerge of the elements in the state space, which can be transformed to real space. Movements and velocities are now known and also the complete set of driving and loading forces of the jaw system.

Unfortunately, even in a system with few degrees of freedom, the computational task is large, since, for every time step Δt, a full optimisation in state space needs to be made, which, using the polytope method, requires the evaluation of the energy function about $10n^2$ times, in which n is the number of degrees of freedom of the mechanism. However, with modern computers, this task is well balanced by the advantage of biological realism in the muscle model, compared to analytical models with simple spring-like muscles. If the nervous system recruits the muscles in such a way that the energy minimum goes through an adequate closed trajectory in the state space of the mechanism, a regular chewing movement is performed.

An interesting problem emerges here: if a local energy minimum in which the jaw system resides does not move away in state space, but vanishes by becoming more and more shallow, the jaw system can be driven to a sudden movement. In the analogy of Fig. 2, the ball will move suddenly to the nearest minimum once the one it was lying in has disappeared. This phenomenon is well described by Thom (1975) as one of the seven possible mathematical instabilities or catastrophies that can occur in four dimensions (three spatial ones and one temporal).

The question for motor control of the lower jaw system is: does this jump have to be avoided? During sudden movement, the jaw is hard to control, but the instability could be used for fast movements. The accelerations achieved are not limited by the dynamics of muscle recruitment, since muscles can be recruited before the movement takes place, so that quite substantial energy can be built up in the system. These are welcome features, especially in prey capture. This possibility could be considered when analysing electromyographs (EMGs) from animals that capture prey by explosive movements.

One catastrophe (of Thom's type) cannot be avoided. This is the one which occurs when food breaks. Here the animal usually is unable to predict the exact moment of breaking of brittle food. A sudden jump in state space cannot be avoided. This jump will be accompanied by a sudden transfer of muscle strain energy into the irreversible potential energy which holds the particles of brittle food together. The bonds between these particles are very strong, but have a limited range, so that the only way to overcome them is to apply high shearing forces. Figure 3 illustrates the energetics of this catastrophe. While the closing muscles are recruited at an increasing level, the jaw remains fixed in a local minimum in total energy created by the food. When the applied force is

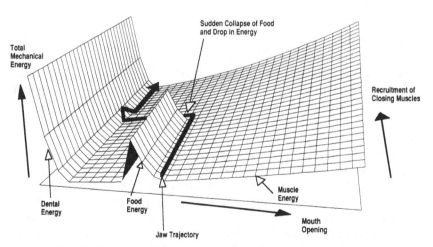

Fig. 3. The total mechanical energy of the jaw system as a function of mouth opening and recruitment of the closing muscles. The food held between the teeth suddenly collapses with increasing recruitment of the muscles. The mouth closes suddenly and total mechanical energy drops.

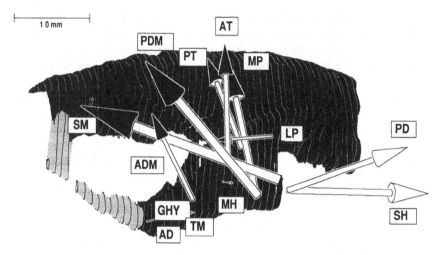

Fig. 4. A three-dimensional reconstruction of the skull, lower jaw and teeth of a rat, together with the maximal forces of the jaw and hyoid muscles. SM, superficial masseter muscle; PDM, posterior deep masseter muscle; PT, posterior temporal muscle; AT, anterior temporal muscle; MP, medial pterygoid muscle; LP, lateral pterygoid muscle; MH, mylohyoid muscle; ADM, anterior deep masseter muscle; GHY, geniohyoid muscle; TM, transverse mandibular muscle; AD, anterior digastric muscle; PD, posterior digastric muscle; SH, sternohyoid muscle.

high enough this energy barrier disappears and the jaw is allowed to close until it hits the energy wall created by the elasticity of the teeth and molars with their suspensions. The nervous system cannot use feedback to control the sudden acceleration of the jaw, because of the limited amount of time. In this situation, the force–velocity relation of co-contracting muscles can be useful to avoid damage of the dental elements. This will be expanded below.

Figure 4 shows the reconstructed skull and lower jaw of a rat. (This reconstruction was made with MacReco, a three-dimensional reconstruction package for Apple Macintosh computers written by Otten and van Leeuwen (1989).) The maximal forces of 13 muscle units, on one side of the head, that have been used in simulations are indicated by arrows. There are more than twice the number of muscles as there are degrees of freedom, which is a common feature in muscle–bone–ligament systems. There are several explanations for this apparent redundancy of muscles. First, the degrees of freedom are not the only properties of the mechanism to be controlled. Stiffness has also to be controlled. Co-contraction of antagonists is a common strategy for stiffness control. Humans who bite on a force transducer at increasing force level will use co-contraction of opening and closing muscles of the mouth when they know the transducer may suddenly yield at a pre-set force level (Miles & Wilkinson, 1982). But antagonistic co-contraction cannot change the direction of maximal stiffness. More than two muscles per degree of freedom are needed to do this. Second, muscles tend to specialise in either maximal force or maximal extension range. Given a certain muscle volume, these two demands are mutually exclusive and therefore two muscles may span the same degree of freedom in the same direction. Third, organisms are not designed the way machines are designed. In evolution any advantage that turns up, even by accident, can be used and so distributed solutions evolve (many muscles across a joint) for problems for which machines would have concentrated solutions (only one engine in an automobile). Clearly, distributed solutions are less vulnerable when parts are damaged. Part of the function is always maintained.

In order to discover the control strategy used for chewing in the rat, EMG signals of all muscles were recorded. All relevant muscle parameters were measured from reconstructions from serial sections. The properties of the food (rat pellets) were determined after stealing it from the mouth of a rat. This was done because saliva changes the mechanical properties of the food. Simulations of chewing were performed on a fast microcomputer (Apple Macintosh IIx).

The calculated chewing movements look very much like real chewing movements and comparison with movement analysis performed by Weijs & Dantuma (1975) showed a good numerical fit (Otten, 1987b). This showed that the arrangement of nested models – with its transformations from architecture to functional characteristics – is useful. The compound model can now be used to evaluate the meaning of recruitment patterns and of physiological muscular properties.

The functional meaning of muscle force–velocity relations were evaluated by leaving them out of the model, an operation which can never be performed experimentally. It appeared that the force–velocity relation in muscles provides a mechanism for dealing with unexpected events, like breaking of food while chewing (Otten, 1987b). This is because at sudden accelerations of the lower jaw when food breaks, the opening muscles (which are slightly active) are stretched suddenly and therefore increase their force due to the force–velocity relation before any control by the nervous system can occur. This effect is shown in Fig. 5. The feed–back loop of reflex control is far too slow to prevent any damage to the oral structures, but the build up of force due to stretch of an active opening muscle has the same short time constant as the formation of cross-

(a)

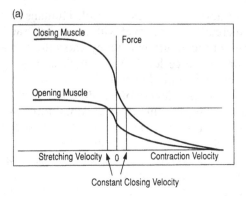

(b)

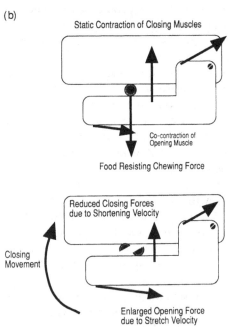

Fig. 5. (a) The force velocity relations of opening and closing muscles guarantee a limited closing velocity of the mouth before feedback occurs. (b) The force balance during static biting and after collapse of the food.

bridges. This is an essential mechanism, especially in small animals with well-developed muscles, because the inertia of the lower jaw and mandibular muscles is low. The rat for instance can produce about 100 N of biting force, while the weight of the lower jaw plus half the weight of the attached muscles is about 4 g. The resulting acceleration is so high that, when food breaks, the lower jaw would hit the skull at a speed of about 16 ms^{-1} reached in a period of 0.6 ms.

From the EMG patterns it appeared that some muscles have a double activation peak during one full chewing cycle. These peaks occur during the power stroke and during mouth opening. The muscles that show this behaviour are the mouth opening muscles (which is to be expected from the principle of co-contraction outlined above), the two lower jaw protruding muscles (the lateral and medial pterygoids) and the transverse mandibular muscle. An animation sequence on the computer produced by the myocybernetic jaw model suggested an explanation for the double-peaked activity of protrusion muscles. The first peak of activity adds to the power stroke, since the rat uses mainly protrusion during the power stroke and uses hardly any lateral movements of the lower jaw. The explanation of the second peak is less straightforward: the protrusion muscles are active during mouth opening to prevent the lower jaw from moving caudad, since the opening muscles pull mainly in a caudal direction. If the lower jaw moved caudad during mouth opening, it would be dragged through the food rather than pulled out of the food, after which a closing movement could be performed. An unexpected member of the double-peaked activity group of muscles is the transverse mandibular muscle, which connects both the lower jaw rami just rostrad from the molars. Humans and other mammals that have ossified symphyses do not have this muscle, but the rat has articulation between the rami of the lower jaw. By letting only the transverse mandibular muscle be active in the myocybernetic model, an interesting function became apparent: protrusion of the lower jaw. Figure 6 explains this: since the fossae of the temporomandibular joint on the skull are not parallel but converge slightly in a rostral direction, a force that squeezes the lower jaw rami together tends to make the lower jaw protrude. As the transverse mandibular muscle is a protrusion muscle, the explanation for the double-peaked activity also holds for the transverse mandibular muscle.

The direction of the sum of all maximal forces produced by the muscles attached to the lower jaw is rostro-dorsad. This is not surprising considering the need for high forces at the lower front elements during the chiselling actions performed on hard food. The same direction is needed during the chewing power stroke. Two strong muscles are attached to the hyoid that pull in the caudal direction: the posterior digastric muscle and the sternohyoid muscle. Animation produced by the myocybernetic model suggested one of their functions. During chewing the hyoid moves through an almost elliptic trajectory with some phase advance relative to the rostral tip of the lower jaw. This phase advance is driven caudally by

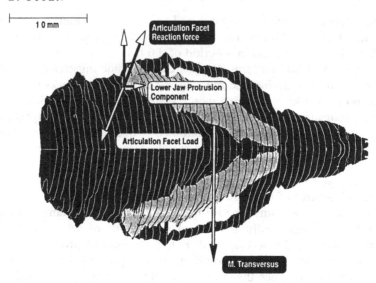

Fig. 6. A three-dimensional reconstruction (ventral view) of the skull and lower jaw of a rat with an analysis of the force generated by the transverse mandibular muscle. Decomposition of the load on the fossa of the temporomandibular joint on the skull results in a protrusion component of the lower jaw, M, musculus.

the posterior digastric muscle and the sternohyoid muscle. During the caudad movement of the hyoid, the lower jaw is not yet opening, so the anterior digastric muscle and the mylohyoid muscle are stretched considerably while active (eccentric contraction), so that their passive components help in opening the lower jaw.

Muscle spindles

Up to this point, the control signals in the motor pattern are taken as given input to the periphery and the movements and forces are calculated as the behaviour of the periphery. In investigating motor control, however, sensory information from the periphery may be highly relevant. Therefore, it is useful to formulate models of proprioceptors, which calculate firing rates from mechanical events. The most intriguing type of proprioceptor is the muscle spindle, which is a modified muscle fibre with its own afferent and efferent (gamma) innervation. Muscle spindles have attracted a great deal of attention and research over the last decade. Their functionality is well described, but proved to be hard to explain.

A model simulating the mechanical events inside muscle spindles was formulated (Schaafsma, Otten & Van Willigen, 1990). The control parameters were the gamma drive (of both the static and the dynamic nuclear bag fibres) and the external length. The mechanical parameters were the stiffness and (linear) damping characteristics of muscular and sensory elements of the intrafusal fibres. This muscle spindle model gave a reasonable explanation of the high sensitivity of muscle spindles to small movements; the short-range stiffness (or stiction) of the non-sensory part of the intrafusal muscles fibre (short-lasting cross-bridge attachments that give way under stretch) appeared to cause a quick stretch of the sensory region at the onset of a muscle elongation. The force–velocity curve of the activated intrafusal muscle fibre added to this effect. A muscle spindle translates changes in muscle length in a highly non-linear way (which is one of the reasons for building a muscle-spindle model). Since in the myocybernetic model all muscle lengths are known during movement, the mechanical inputs to the muscle spindles are also known. This gives the possibility of calculating the afferent signals to the nervous system, which opens the way for understanding motor learning.

The other type of input to the muscle spindle (see Fig. 7), the gamma

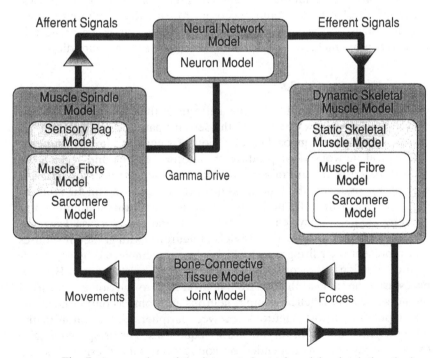

Fig. 7. An overview of the integrated set of models, simulating the jaw system. Note the distributed, hierarchical and directional organisation of the models.

drive, partly runs in synchrony with the alpha motor neurons, but may have its own course during changes in motor strategy. The function of gamma drive may become clear with the ensemble of models when indications about the type of information that the nervous system needs in programming its motor patterns becomes available.

In the jaw system of the rat, muscle spindles do not occur or are very rare in the digastric muscles (both anterior and posterior) and the pterygoid muscles (both lateral and medial) (Rokx, Van Willigen & Jansen, 1984). This same group of muscles shows double peaked activity during the chewing cycle. During closing movements of the jaw, there is no sensory information coming from protrusion and opening muscles. One could argue that the jaw system is overdetermined in the sense that what happens in the closing muscles is a complement of what happens in the opening muscles, so only one set of muscles needs to be equipped with sensors. An indication of this is that protrusion and opening muscles act as pure effectors, which are controlled by proprioceptive activity of the jaw-closing muscles (Jüch et al., 1984; Van Willigen et al., 1986). But usually distributed solutions are found and sensors appear in all muscles. The high-threshold periodontal receptors are able to contribute to a jaw-opening digastric reflex (Dessem et al., 1988), so one could argue that these receptors make muscle spindles in the jaw openers superfluous.

Neural networks

Once one has established the motor patterns of the chewing muscles, the movements of the jaw system and the sensory patterns coming from the muscle spindles, the myocybernetic loop of signals is almost complete. The neural networks that produce the motor patterns and receive the sensory patterns are still missing and one may wonder just how much is missing for understanding motor control of the jaw system. In order to improve insight into the problem of pattern transformation and generation by the nervous system, a model of a neural network was designed. The network was composed of models of neurons (with graded responses that represent their firing rates) which were interconnected by asymmetrical connections (the connection from neuron A on to neuron B is not necessarily the same as the one from neuron B on to A). Some neurons of the network were labelled as input neurons and some as output neurons. The input (or sensory) neurons received peripheral information in the form of spike trains with varying firing frequencies. The firing frequencies of the output neurons depended on connections with interneurons and the sensory neurons. The connections had delays, while the neurons had short memories for their previous firing rates. In contrast to pattern recognition, in motor control the input to a neural network depends strongly on its output, since the magnitude of the signals from peripheral receptors to the network depends on the level of activity of the motor neurons which cause the movement and excite the proprioceptors. Also, such an active motor network never reaches a steady-state output pattern of activity, because it is constantly driven by the changing input.

From simulations, it became apparent that neural motor networks, composed of the same number of neurons as the number of space-time units needed to describe a motor pattern, were able to learn to produce at least two different motor patterns from two totally unrelated sensory patterns. The learning process was simulated by selecting randomly a possible connection between any pair of neurons, making the connection stronger or weaker and letting the newly formed configuration perform the required transformation from sensory pattern to motor pattern. If the motor pattern appeared to improve or remained the same, the change in connection was maintained. If the motor pattern became worse (differed more from the ideal than before the synaptic change was introduced), the connection was reset to its former state.

It should be stressed that this type of learning can never happen in real life, since a so-called 'supervisor' is required which passes the information about the success of a change in a synapse back to the synapse. Such a supervisor is not known to exist in organisms. This type of learning can be used to reveal only the potential calculating power of a network. It was interesting to see that, if the same pattern transformation was requested several times, different solutions were found each time. Naturally the interneurons behaved very differently in the various solutions, but the required motor pattern was always the same. There is a high amount of connection redundancy in terms of the pattern transforming function of the network.

Another striking finding was that such networks have strong pattern interpolating properties: if two different motor output patterns were 'learned' by a network, any sensory pattern falling between the ones that were used for learning was translated by the network into a motor pattern that was in-between the two initial motor patterns. This is a very important emerging property of a neural motor network, since it implies that from a limited number of learned motor patterns an almost infinite set of patterns can be interpolated to deal with completely new circumstances. For chewing such circumstances may be the texture of food, which has a large range. By learning to handle just a few types of texture, a large range of textures can be interpolated by the network.

At present it is believed that rhythmic movements in vertebrates are driven by so-called central pattern generators (Cohen, Rossignol &

Grillner, 1987), although the strong influence of proprioceptors is acknowledged (Jüch *et al.*, 1985). In order to investigate whether a simple neural network, with only a few connections given beforehand, has the potential to form a pattern generator, a learning rule was applied to such a network. This learning role is: neurons that have *low* average firing rates will tend to *increase* their sensitivity for synaptic input, whereas neurons with *high* average firing rates will tend to *decrease* their sensitivity. This rule, that of controlled sensitivity, tends to suppress any local pattern of activity in a network. It has the same effect on the neural pattern of activity as the contrast control on a television monitor.

The formation of a central pattern generator was simulated in a neural network consisting of a matrix of 15 by 15 neurons. Each neuron in the matrix could make connections with the 24 closest neurons. No connections were made at the start of the simulation, except for one single string of exciting connections between one group of neurons in the matrix. This string of connections resulted in a static pattern of neural activity in the matrix: only the string had higher firing rates; the other neurons fired at a lower spontaneous firing rate. The controlled-sensitivity rule was applied until the static pattern disappeared. This happened because the rule formed new connections that tended to reduce neural patterns. At the point of disappearance of the pattern an oscillation emerged, sometimes with quite a complex time pattern. The oscillation frequency of the pattern generator could be controlled by letting neurons, which were not part of the oscillating group but projected on to it with an inhibiting connection, be more or less active. The shape of the cyclic pattern of activity could also be regulated by neurons projecting with both exciting and inhibiting connections on to the loop of neurons.

Perhaps this simulation illustrates the formation of a central pattern generator for chewing or any other form of cyclic activity. It has some features that can be recognised in an organism, such as controllability of both rate and shape of the pattern and the way it emerges from a simple lay-out of connections, which is known from ontogenetic studies of the nervous system. The interpolating property of neural nets for learned pattern transformations suggests a solution for the problem of needing an infinite set of neural nets for an infinite variety of tasks. The simulations are, however, still far away from being fed with real data from neuro-anatomy and neurophysiology. The reason is that both fields tend to hide any explanation of neural function of more than a dozen neurons in a fog of details. Neural network studies are still in their infancy, trying to discover the basic concepts of neural function. The field has certainly produced a number of astonishing features of neural networks, especially when biologically realistic assumptions are used, but it can not yet handle the tremendous flow of data coming from neuroanatomy and neurophysiology. In order to perform any rigorous tests on neural network models, these data will have to be handled in the near future.

Discussion

The orchestration of muscle activity during feeding is one of many problems in motor control. The jaw mechanism in mammals is not an average motor system, since due to required high static forces the system is stiff in dynamic circumstances. Also there are no muscles that span more than one joint and so the whole multi-joint planning problem of muscle recruitment, well known from studies of human arm control, is absent in the jaw system. There is, however, no system more suitable for studying the problem of open loop control and prevention of damage.

The success of predicting movements and forces of the jaw system of the rat from recruitment patterns of the muscles is largely due to the fact that the components of the myocybernetic model were tested separately. Figure 7 shows the complete set of models. The muscle spindle model has been tested also, but not in the context of the myocybernetic model. The same holds for the neural network models. That certainly is an unfinished aspect of the modelling of the control system of the lower jaw and has much to do with blank areas in the map of knowledge concerning the nervous system and its interfaces with the periphery.

Modelling is one way of analysing the peculiarities of a complex dynamical system. Its value, however, depends completely on the experimental data that can be used to give suggestions for the structure of models, and then to feed the models and to test them. Large quantities of experimental data do not guarantee explanations, especially if they have been gathered only because some device was available that gave the opportunity to take measurements. On the other hand, there are so many languages in mathematics that it is easy to find one that resonates with nature. It is, however, more difficult to find the mathematical language that not only describes nature well but makes correct predictions about properties yet to be measured. Clearly this can be done only when experimental and theoretical science are thoroughly integrated.

Acknowledgments

Jan Douwe van Willigen is thanked for his linguistic revision of the chapter. Many of the ideas presented here were handled by him in an early phase with enthusiasm and erudition.

References

Akster, H.A., Granzier, H.L.M. & ter Keurs, H.E.D.J. (1984). Force–sarcomere length relations vary with thin filament length in muscle fibres of the perch (*Perca fluviatilis* L.). *Journal of Physiology*, **353**, 61P.

Alexander, R. McN. (1968). *Animal Mechanics*. Sidgwick and Jackson, London.

Arnold, V.I. (1978). *Mathematical Methods of Classical Mechanics*. Springer-Verlag, New York.

Aubert, X. (1956). *Le Couplage Energetique de la Contraction Musculaire*. Thesis, Editions Arscia, Brussels.

Benninghoff, A. & Rollhäuser, H. (1952). Zur inneren Mechanik des gefiederten Muskels. *Pflügers Archive für die gesampte Physiologie des Menschen und der Tiere*, **254**, 527–48.

Cohen, A.H., Rissignol, S. & Grillner, S. (eds.) (1987). *Neural Control of Rhythmic Movements in Vertebrates*. John Wiley & Sons, New York.

Dessem, D., Iyadurai, D.O. & Taylor, A. (1988). The role of periodontal receptors in the jaw-opening reflex of the cat. *Journal of Physiology*, **406**, 315–30.

Gans, C. (1982). Fiber architecture and muscle function. In *Exercise and Sport Sciences Reviews*, vol. 10, ed. R.L. Teijung, pp. 160–207. Franklin Institute Press, Philadelphia.

Gans, C. & Bock, W.J. (1965). The functional significance of muscle architecture – a theoretical analysis. *Ergebnisse der Anatomie und Entwicklungsgeschichte*, **38**, 115–42.

Gaspard, M. (1965). Introduction l'analyse bio-mathematique de l'architecture des muscles. *Extrait des Archives d'Anatomie, d'Histologie et d'Embryologie normale*, **XLVIII**, 95–146.

Hatze, H.A. (1978). A general myocybernetic control model of skeletal muscle. *Biological Cybernetics*, **28**, 143–57.

Hill, A.V. (1938). The heat of shortening and the dynamic constants of muscle. *Proceedings of the Royal Society, Ser. B*, **126**, 136–95.

Heukelom, B., van der Stelt, A. & Diegenbach, P.C. (1979). A simple anatomical model of muscle and the effects of internal pressure. *Bulletin of Mathematical Biology*, **41**, 791–802.

Jüch, P.J.W., van Willigen, J.D., Broekhuijsen, M.L. & Ballintijn, C.M. (1984). Masseter, digastric and omohyoidal responses from weak mechanical oral stimuli in the chewing rat. *Brain, Behaviour and Evolution*, **25**, 166–74.

Jüch, P.J.W., van Willigen, J.D., Broekhuijsen, M.L. & Ballintijn, C.M. (1985). Peripheral influences on the central pattern-rhythm generator for tongue movements in the rat. *Archives of Oral Biology*, **30**, 415–21.

Miles, T.S. & Wilkinson, T.M. (1982). Limitation of jaw movement by antagonist muscle stiffness during unloading of human jaw closing muscles. *Experimental Brain Research*, **46**, 305–10.

Nelder, J.A. & Mead, R. (1965). A simplex method for function minimization. *Computer Journal*, **7**, 308–13.

Otten, E. (1987*a*). Optimal design of vertebrate and insect sarcomeres. *Journal of Morphology*, **191**, 49–62.

Otten, E. (1987*b*). A myocybernetic model of the jaw system of the rat. *Journal of Neuroscience Methods*, **21**, 287–302.

Otten, E. (1988). Concepts and models of functional architecture in skeletal muscle. In *Exercise and Sport Sciences Reviews*, vol. 16, ed. K.B. Pandolf, pp. 89–139. Macmillan, New York.

Otten, E. & van Leeuwen, J. (1989). MacReco, a 3-D reconstruction package of serial images. *European Journal of Cell Biology*, **25** (48), 73–7.

Pfuhl, W. (1937). Die gefiederten Muskeln, ihre Form und ihre Wirkungsweise. *Zeitschrift für Anatamie und Entwicklunsgeschichte*, **106**, 749–69.

Pollack, G.H. & Sugi, H. (eds.) (1984). *Contractile Mechanisms in Muscle*. Plenum Press, New York & London.

Rokx, J.T., Van Willigen, J.D. & Jansen, H.W.B. (1984). Muscle fibre types and muscle spindles in the jaw musculature of the rat. *Archives of Oral Biology*, **29**, 25–31.

Schaafsma, A., Otten, E. & Van Willigen, J.D. (1990). A model for muscle spindle primary firing. 1. Linear mechanics and stiction. *Journal of Neurophysiology*, in press.

Scherz, G. (1963). *Pioner der Wissenschaft. Niels Stensen in seinen Schriften*. Munksgaard, Copenhagen.

Thom, R. (1975). *Structural Stability and Morphogenesis*. W.A. Benjamin Inc., Reading, Ma.

van Willigen, J.D., Jüch, P.J.W., Ballintijn, C.M. & Broekhuisen, M.L. (1986). A hierarchy of neural control of mastication in the rat. *Neuroscience*, **19**, 447–55.

Wallinga-DeJonge, W. (1980). *Force Development in Rat Skeletal Muscle: Measurements and Modelling*. Thesis, Enschede, The Netherlands.

Weijs, W.A. & Dantuma, R. (1975). Electromyography and mechanics of mastication in the albino rat. *Journal of Morphology*, **146**, 1–34.

Woittiez, R.D., Huijing, P.A., Boom, H.B.K. & Rozendal, R.H. (1984). A three-dimensional muscle model: a quantified relation between form and function of skeletal muscles. *Journal of Morphology*, **182**, 95–113.

M. R. HEATH

The basic mechanics of mastication: man's adaptive success

Man is an omnivore: the uniquely adapted animal with a mouth which was well suited to a wide diet even before the advent of cooking. Manual collection of food allows a shorter snout which improves the mechanics of oral processing of food; this in turn allows a wider natural diet by inclusion of foods that would be impervious to enzyme action in the human gut without mastication. Both the reduction in particle size and wetting with saliva accelerate enzyme action in the gut and create a swallowable bolus. The success of this reflex activity is indicated by the infrequency of any consciousness of the process despite the continuing generous subconscious control.

The reduction of particle size is fundamentally important in the oral breakdown of natural unprocessed foods. Although foods developed by industry bear little structural relation to natural foods, basic appetites persist and the control of mastication maintains many general features used by other mammals. One aspect of human chewing behaviour is that much pleasure ensues from taste and smell. The rate of release of taste may be a governing factor in the popularity of many foods and may have unconsciously been important in their 'design'. Taste is elicited by substances passing into solution in the saliva (e.g. the sugar from chewing gum, toffee), some foods are first hydrated by saliva (e.g. potato crisps, biscuits, breakfast cereals), others dissolve or melt (e.g. jellies). These events all involve continuously changing the internal characteristics of the foods to achieve a continuous source of taste and smell for the duration of chewing. Melting and significant hydration probably occur rarely in unprocessed foods.

In foods which possess a high moisture content, the expression of moisture may release the taste. An obvious class of natural food which behaves this way is juicy fruit. Moisture forced out by bursting turgid cells is probably far more significant than any particle size reduction. Critical for the release of moisture from foods in this manner is the position of the packets of moisture within the food, their individual size and their ease of release. The noise of fracture of crisp vegetables such as celery is probably associated with freshness and perceptions of noise, texture and taste are interrelated in food selection.

Studies on jaw movement indicate that chewing involves perception of the food being chewed and therefore fundamental studies of the process should provide an understanding of the sensory basis for control. There are two processes involved in the fracture of foods:

1. Selection by the tongue and cheek of particles into a bolus between the teeth giving rise to a *selection function* for the probability of fracture.
2. A *breakage function* representing the extent of fracture of the selected particles (Epstein, 1947).

Numerical solutions for these functions have been achieved with computers (Gardner & Austin, 1962; Lucas & Luke, 1983b; Lucas et al., 1986; Olthoff, 1986; Voon et al., 1986).

A review of the phases of mastication illustrates how the human mouth is morphologically and structurally adapted for coping with diverse foods and even adapting to loss of teeth and ageing.

Oral morphology and the control of food

Man's short snout is distinctive. While many explanations have been offered for facial shortening, notably the improvement in biting forces produced by the masticatory muscles (Preuschoft, 1989), the relevance of morphological change to the control of the food bolus by facial muscles seems to have been neglected. The short snout is fundamental to the action of the lips for careful oral assessment of foods and facilitates control of particles during the early cycles prior to bolus formation (Fig. 1).

The orbicularis oris muscles, the principal muscles of the lips, have mobile origins from *muscles* at the corners of the mouth whose anatomy and action will therefore be described first. Their action and that of other facial muscles was described in detail by Lightoller (1925). He named the knot of muscles which converge behind the corner of the mouth the *modiolus* (like the nave of a spoked wheel) (Fig. 2). The peculiar value of this structure for food control was identified by Fish (1934). The morphological importance of the near vertical muscles is that they form a curtain which can control food on to the occlusal table. Differential activity of the elevator muscles allows fine control of the buccolingual position at which pressure is placed on the food (Fig. 3). The origins of these muscles on the maxilla give different lines of action so that coordinated activity of

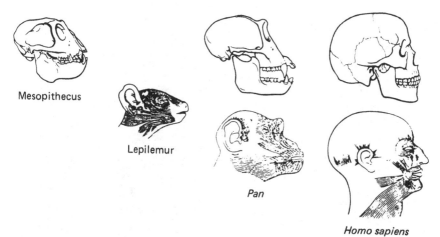

Fig. 1. Man's short snout and related musculature contrasted with *Mesopithicus*, an anthropoid from the Miocene; *Lepilemur* (lemur) a prosimian which shows incipient primate differentiation of facial muscles and *Pan* (chimpanzee) whose lips grasp food and move vertically in facial expression. (Redrawn from Young, 1971, pp. 420 and 434.)

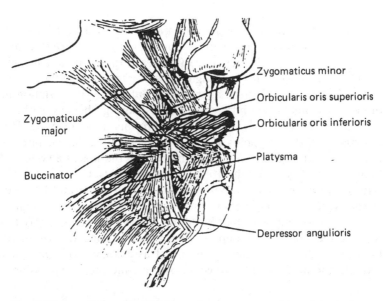

Fig. 2. The modiolus: the facial muscles converge and intermingle forming a dense knot to provide a mobile origin for labial muscles. (Redrawn from Lightoller, 1925.)

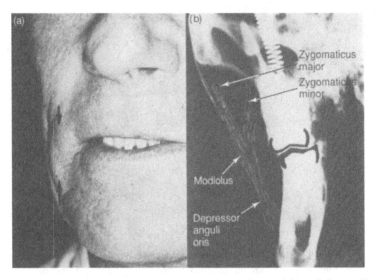

Fig. 3. (a) The cheek viewed obliquely shows the position of the modiolus. (b) A radiograph of a dry skull from a similar angle. The line of
action of the elevator muscles zygomaticus major and minor and depressor anguli oris are added diagrammatically.

elevator and depressor muscles will apply tension, holding a food bolus
between the teeth. A finger inside the corner of the mouth can readily
feel pressure created by activity. In contrast, the zygomaticus major will
maintain the tension in a less lingual position to preclude the cheek being
bitten. Their lines of action enable this control despite variation in gape
or lateral mandibular shift which occurs during chewing with forward
translation of the contralateral temporo-mandibular joint. The oral
mucosa over the modiolus is firmly attached by collagen fibres to the
epimysium. Its action can be seen by parting the lips as a sweet is bitten
between the cheek teeth. Reflex activity holds the sweet, balancing
outward pressure by the tongue until the teeth penetrate the food.

The orbicularis oris muscles of the lips act by tensing the lips against the
anterior teeth (Fig. 4): small facial muscles act perpendicularly to effect
detailed movement of the lips so that, acting in concert with the tongue,
they can control food during its initial assessment and, *inter alia*, can
provide fine movements involved in speech and facial expression.

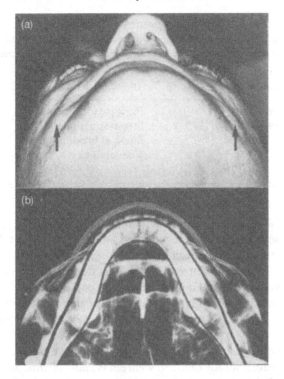

Fig. 4. (a) The curvature of the lips, seen from beneath, between their origins from the modioli. (b) A radiograph of a dry skull from a similar angle with modiolar and labial muscles drawn diagrammatically. The curvature contrasts with other hominids.

Initial assessment of food

Familiar foods are selected by sight and smell but external food properties can be perceived orally by the lips, the tip of the tongue and the incisor teeth. Subsequently internal physical properties are detected during incision and mastication. The difficulty of characterising foods, discussed by Szczesniak (1963), justifies separate consideration of perception of their external and internal physical properties (Table 1).

The *external characteristics* of foods (size, shape, stickiness, roughness, etc.) are detected by the tongue, lips and incisors as the food is held and bitten. The oral mucosae are liberally innervated with sensory nerve endings which lie beneath the epithelium (Dixon, 1963). The lips are particularly sensitive to temperature, while the ability to make the finest two-point discrimination (1–2 mm) is found on the upper surface of the

Table 1. *Physical properties of foods*
A wide range of properties contribute to the oral assessment of food

External (surface features)	Internal (not expressed at surface)
Particle size	Fracture toughness
Particle shape	Stress–strain relationship
Number of particles	Stress at breaking strain
Stickiness	Viscoelasticity
Roughness	

tip of the tongue (Ringel & Ewanowski, 1965; Laine & Siirila, 1971). Superficial lingual sensory nerves are very fast adapting. Light mechanical stimulation produces a short discharge of pulses but none thereafter despite continued stimulation (Porter, 1966). This fast adaptation serves the sensory perception of lingual contact that is essential during chewing and speech. The incisors are capable of extremely fine discrimination. Single neurons have thresholds to displacements of 2–3 μm (reviewed by Anderson, Hannam & Matthews, 1970), but for dynamic detection of foils between natural incisors the average threshold is 8–12 μm (Utz, 1983) whereas the threshold for denture wearers is 200–300 μm, about 25 times greater (Utz & Wegmann, 1985).

The discrimination of *internal texture* of food demands more complex interpretation. The inclusion of very hard particles mixed within softer particles involves assessment of features of both. Owall & Vorwerk (1974) showed that small steel balls inserted into peanuts could be detected down to 0.4 mm diameter by many dentate individuals. However, denture wearers were dramatically poorer in their discrimination, indicating the role of the periodontal membrane for such fine force/displacement discrimination. The complexity of this task can be inferred from the very much finer discrimination of minute aluminium oxide particles suspended within yoghurt – the absolute threshold being about 15 μm (Utz, 1983); the contrast between the yoghurt vehicle and the particles allows them to be detected at a low force : displacement ratio of periodontal fibres.

The assessment of internal texture also contributes to the initial acceptance of unfamiliar foods as can be inferred from the significantly slower chewing speeds used by subjects in Owall's (1978) experiments. They chewed slowly when they knew that peanuts might enclose a steel ball – a proper caution which may explain the slower closing speed in early cycles of chewing sequences with unfamiliar foods (E. Kazazoglu &

M.R. Heath, unpublished data). The more familiar the food, the less initial assessment is needed. One can infer that the ability to perform initial assessment allows exploration of a wide diet.

Patterns of movement

The process of mastication is basically achieved by movement of the jaws, the voluminous literature on which was reviewed by Bates, Stafford & Harrison (1975*a,b*) and more recently by Jemt (1984). Although little of this research covered the relevance of movements to the mechanics of food comminution, some implications are worth considering.

Jaw movements are commonly viewed in two planes. From the side, movements are very repeatable in form and vary little with the type of food that is chewed (Fig. 5). From the front, however, jaw movements appear to be very different. They are variable even within one sequence, and opening movements follow a path different from that of closing movements (Fig. 6; and Heath & Lucas, 1988). The opening path is

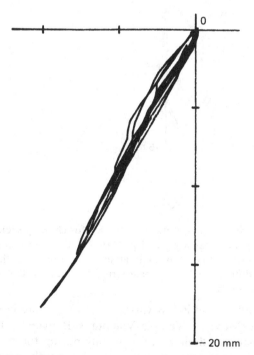

Fig. 5. Jaw movements viewed from the side: in this plane the pattern of chewing movements varies little.

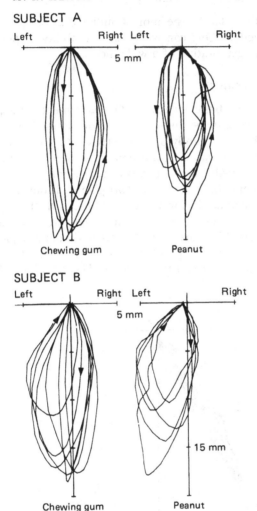

Fig. 6. Variations in patterns of movement used for chewing seen from the front. Each person has a general pattern – note that steep molar guidance restricts subject B's movement at and near contact. Within the general pattern differences occur, most strikingly irregularities during fracture of brittle foods.

usually closer to the midline and is less variable than the more laterally positioned closing path (Mongini, Tempia-Valenta & Benvegnu, 1986). In fact anatomical constraints limit movement only during the tiny but very important part of the cycle when the teeth are in or nearly in contact (Wickwire *et al.*, 1981).

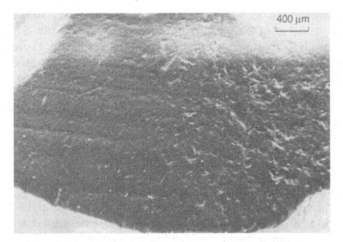

Fig. 7. Functional abrasion of an acrylic denture tooth showing two facets; the left-hand zone records the direction of articulation between opposing teeth; the other shows pitting caused by abrasive food.

The striking feature common to all studies has been repeatability of the general pattern exhibited by each person (Jemt & Hedegard, 1982). This is particularly true with homogeneous viscous food such as chewing gum. Comparisons between different persons chewing a range of foods suggest that the variation between people is larger than that between foods (Fig. 6). Individual patterns are thought to become established early in life (Ahlgren, 1966) but their modification in response to change in dental state justifies further detailed longitudinal study. The wear of acrylic teeth on dentures can indicate which teeth are used for chewing abrasive foods and even, in some cases, the direction of jaw movement (Heath, 1986) (Fig. 7).

The variability of individual cycles for particulate foods shows clearly that the basic pattern can be modulated, presumably related to variation in features of the bolus for each stroke (Fig. 6). It therefore follows that recordings which average out the movement between sequences will lose any relationships that might indicate textural perception. The responsiveness of movement patterns to oral stimuli is also clearly shown by the increase in vertical movement with increased volume of peanuts in experimental mouthfuls (Lucas *et al.*, 1986).

The underlying patterns of movement for each individual are easily recognised during the kneading phase; some within-subject variation occurs with different foods and different mouthful sizes, but more striking variation occurs between subjects. The most closed part of these cycles can be related to the occlusal form but the overall pattern justifies further study. The concept of a central pattern generator within the brain stem which drives this basic individual pattern of movement is now well accepted (Lund & Olsen, 1983). This concept can allow for modulation based on mouthful size, internal texture, etc., as perceived during incising or the first biting stroke so as to provide efficient management of mastication, modulated further as chewing progresses (Taylor, Appenteng & Morimoto, 1981; Otten, this volume).

Mastication: early cycles

One essential action in every masticatory cycle is the placing of food between the occlusal surfaces of the teeth. Following incision, movement of the morsel of food is fast and accurate – movement enabled by the structure of the human tongue, which has no bony skeleton. The intrinsic muscles take origin from collagenous septa and the epimysium of the extrinsic muscles. The range of movements is consequently unrestricted.

During the first few cycles of a chewing sequence, food structure is altered to produce a more or less coherent bolus: brittle structures fracture, fibres are torn and shredded and viscoelastic elements are kneaded and wetted with saliva. These first few cycles are slower than later ones and irregularities of jaw movement can be related to masticatory events.

It is during these early cycles, before a coherent bolus is formed, that the effectiveness with which the food is repositioned between the occlusal surfaces is critical for efficiency. The importance of separating this *selection function* is that impaired coordination of the cheeks and/or tongue in manipulation of food is independent of the subsequent action of teeth closing through that food – the *breakage function*. Both functions contribute separately to masticatory effectiveness (Lucas & Luke, 1983a).

Events during early chewing cycles

The responsiveness of the control of chewing to stimuli within the first few cycles is demonstrated by the speed of reflex inhibition which occurs in most chewing sequences. These cycles show smooth rapid mandibular opening followed by accelerating closure until food contact; deceleration on a hard food is followed by renewed acceleration after fracture of food (Fig. 8). Detailed recording shows that muscle activity increases after food contact until fracture occurs. Some 14 ms later muscle activity is typically inhibited; this inhibition is affected little by anaesthetising the periodontal membrane of the teeth biting into the food, despite the sensitivity of their sensory nerve endings (F. Muller & M.R. Heath,

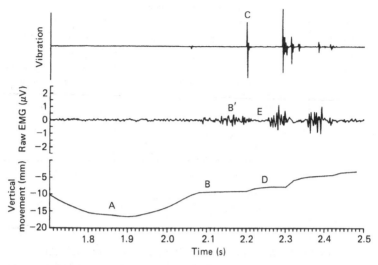

Fig. 8. The responsiveness of masticatory muscles to events during chewing. Vertical mandibular movement from the widest gape (A) shows deceleration with contact on a sweet (B) with increasing muscle activity (B'). At first fracture shown by vibration (C) the mandible accelerates (D) and some 14 ms later muscle activity is inhibited (E). Subsequent fractures cause further inhibition.

unpublished data). This is but one example of the plurality of sensory information used for monitoring oral function. The abrupt unloading of tension in the periodontal membrane and vibration of the relevant teeth are the most obvious effects of food fracture but a number of other sensory nerve endings are stimulated, notably muscle spindles in the elevator muscles which also experience abrupt unloading and some shortening with mandibular movement. Sounds are detected by the cochlea; vibrations are detected at many sites (Drake, 1965). The difficulties of interpreting masticatory forces arise because so many different muscles share the work at different phases of the chewing stroke and food is moved to different teeth between cycles. Furthermore, different foods may be chewed principally by different teeth but it is impossible to state which without careful, elaborate experiments such as those of Tornberg *et al.* (1985). These authors solved the problem by mounting 14 strain gauges on a full arch bridge for one patient who chewed different types of whole meat and processed meat products. The forces used and rates of loading for meats reflected their physical state. Higher forces and slower loading was used with whole meats than for restructured beef, hamburgers and sausages. This subject rated sensory attributes for these

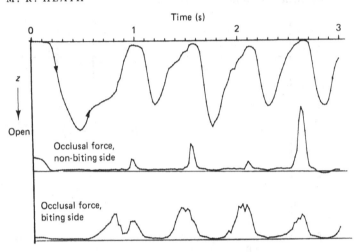

Fig. 9. Force and movement during fracture of a peanut. The first cycle shows two discrete fractures. In subsequent cycles the fractures become progressively contiguous. The first cycle also shows alternate deceleration and acceleration of the mandible which correspond to the chewing side force. (Redrawn from Hsu, 1987.)

meats which correlated impressively with the number of cycles used but not with mechanical bench tests such as the Warner–Bratzler (see Purslow, this volume).

One particularly successful study of forces was that of Bearn (1973), subsequently developed by Hsu (1987), using dentures with a wafer mounting of the teeth so that force was detected from all the cheek teeth on the side being investigated. This demonstrated the collapse of force with each of the many fractures that occur within early strokes when chewing peanuts (Fig. 9).

The kneading or mush phase

Human chewing usually involves an extended sequence of smooth cycles for triturating food. It is striking how soon a repeating cycle of smoother jaw movements is established, typically within a few cycles dependent upon the texture of the food (Fig. 10). Thereafter the bolus is kneaded between the teeth for an extended period, varying between foods and varying markedly between individuals. This extended kneading further reduces particle size, incorporates saliva thus initiating digestion, prepares the bolus to a swallowable consistency and improves the speed of enzyme action in the gut. Although of little nutritional relevance with

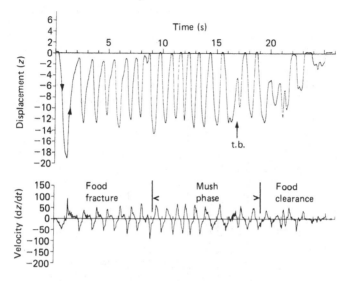

Fig. 10. Vertical (opening) mandibular displacement (z) and velocity (dz/dt) during chewing of a peanut. Multiple fractures occur in the first few cycles, thereafter a regular pattern is used in the mush phase until food clearance. One cycle is used to transfer the bolus from left to right (t.b.).

diets of modern prepared foods, the biological drive to chew retains its satisfaction along with the well-known effect on gastric secretion (Pavlov, 1897).

The precision of the reflex effect of chewing on salivary secretion has only recently been shown. Everyone is conscious of salivary flow in response to psychic and oral stimuli. Kerr (1961) suggested, from one experiment, that parotid salivary flow is restricted to the current chewing side. This was confirmed by Anderson & Hector (1987) (Fig. 11), who further showed that it is the sensation from the periodontal membrane that stimulates this reflex flow. They also demonstrated the apparently linear relationship between activity of the masseter muscle and ipsilateral parotid flow rate (Fig. 12). Intriguingly, loss of all teeth does not obliterate this reflex; in fact the same pattern of unilateral flow has been shown in edentate people (Fig. 13a) but is largely stimulated by sensory endings in the mucosa supporting the lower denture (Fig. 13b; and Heath & Bellwood, 1988). Further, the rate of flow is almost directly proportional to the volume of food being chewed (Fig. 13c).

The parotid gland fails to tire, continuing to respond as long as there are adequate oral stimuli (Garrett, 1987). Saliva in the ductal system is

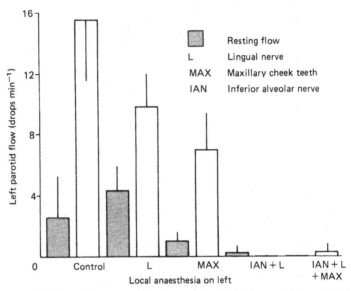

Fig. 11. Parotid salivary flow is stimulated during chewing but this is inhibited by anaesthesia of the periodontal afferent pathways from the relevant teeth. Progressive inhibition is seen as more nerves are anaesthetised. (Data for one subject.) (Redrawn from Anderson & Hector, 1987.)

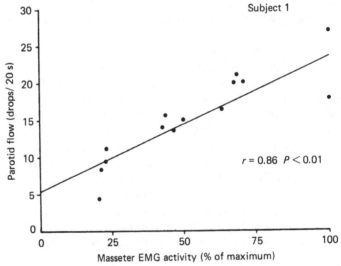

Fig. 12. Salivary flow from the parotid gland increases linearly with ipsilateral masseter EMG activity. (From Anderson & Hector (1987), with permission from the *Journal of Dental Research*.)

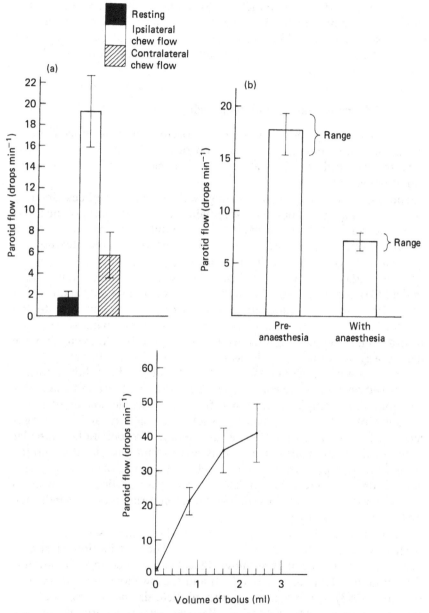

Fig. 13. (a) Denture wearers have a similar masticatory–salivary reflex, despite the loss of all periodontal ligament, showing unilateral parotid flow on the chewing side. (b) Anaesthesia of the mucosa beneath the lower denture halves the flow rate. (c) Flow rate responds to stimuli evoked during chewing on larger volumes of food.

actively expressed by myoepithelial cells so that the flow can be related to stimuli (Anderson, Hector & Linden, 1985). The extensive literature on saliva has been reviewed by Garrett (1987) and the reader is also referred to the published proceedings of the 10th International Congress of Oral Biology (Hand, 1987).

Clearance phase and swallowing

Prior to swallowing, one or more cycles occur that are characteristic of the movements involved in preparation of a swallowable bolus. These have received little previous attention, but some striking differences are found between foods (Fig. 14).

The sequence of movements during swallowing was well described by Adran & Kemp (1955) but interest remains in the consistency of the bolus that is perceived as swallowable because of inferences that can be drawn about the process of mastication and the perception of food texture.

Swallowing is arbitrarily divided into three stages on the basis of the position of the food bolus (Jenkins, 1978). Stage I is in the mouth and is initiated voluntarily. Stage II is in the pharynx and lasts only about 1–2 s. Stage III is in the oesophagus and ends only at the entrance to the stomach. The initiation of a subconscious swallow is obviously triggered by oral stimuli and conditions surrounding Stage I; the oral volume being particularly relevant for fluids, as recently discussed, with new data, by Speirs, Staniforth & Sittampalam (1988). However, for solids, information based on experience concerning Stages II and III also influences this decision. For example, it is known from fluoroscopic studies of radio-opaque tablets that certain materials stick in the pharynx and oesophagus more readily than others (Hey et al., 1982). The peristaltic muscular waves of the pharynx/oesophagus may drive softer particles down to the stomach more rapidly that harder particles. The timing of swallowing may depend on oral perception to select appropriate particle size, wetness and/or plasticity. The effectiveness of this monitoring process is indicated by the rarity of choking.

Dental research has mostly been concerned with the association of food particle size and the decision to swallow. The distribution of size of particles that are normally swallowed has been termed the 'swallowable threshold' by Manly & Braley (1950) and the 'swallowable composition' by Jiffry (1983). It is not clear whether particle size *per se* could be a cue for the initiation of swallowing. Particle size reduction curves are exponential after a small number of chews (about 10) (Lucas & Luke, 1983a; Olthoff et al., 1984; Sheine, 1979). A common feature of such

(a)

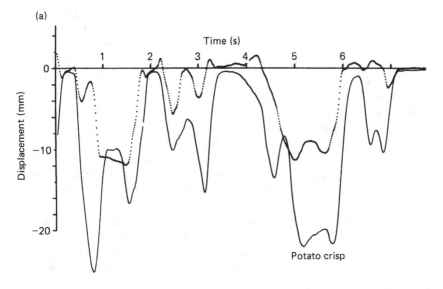

Potato crisp

(b)

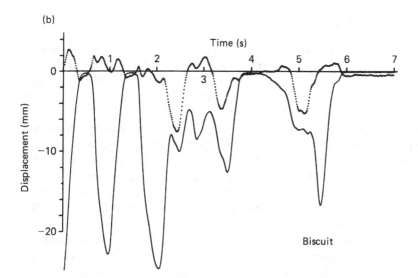

Biscuit

Fig. 14. Oral clearance and swallowing involves mandibular movement so that the tongue can effect clearance. Textural differences are reflected in the duration and range of movements of (a) a potato crisp and (b) a biscuit. Solid line, (z) vertical displacement; dotted line, (x) saggital displacement.

curves is that they do not possess any turning points or inflexions and thus, in themselves, present no feature that could influence the decision to swallow. The physiological limit of perception of particle sizes is far below those sizes that have been found to be swallowable (Utz, 1983; Utz & Wegmann, 1985). The size of particles that are swallowable by a person depends on the volume of food in the mouth, at least for peanuts. Larger particles are swallowed with larger mouthfuls (Jiffry, 1983; Lucas & Luke, 1984; Yurkstas, 1965).

The sizes that are swallowable by different subjects are very variable and may depend on the rate of particle size breakdown that the person can achieve (Dahlberg, 1942; Yurkstas, 1951). The tendency is for people with dentures to chew less well but swallow larger particles rather than fully compensate with more chewing strokes.

Among other factors that might contribute to the decision to swallow are the lubrication of the food (by saliva and expressed moisture) and the intensity (peak or trough) of taste. The rates of hydration, melting and expression of moisture are also likely to follow an exponentially declining form with respect to time.

Effectiveness of mastication

Man's strategies for easy feeding make chewing almost irrelevant to nutrition of young healthy adults; Farrell (1956) showed that unchewed 10 mm cubes of many foods were completely digested. However, this study has never been repeated on elderly people nor on those with any gastrointestinal impairment.

There is a large literature on masticatory effectiveness which was reviewed by Bates, Stafford & Harrison (1976). Not surprisingly there is evidence that some aspects of the dental occlusion are important for effectiveness, but different methods of assessment failed to give good correlations even within studies (Helkimo, Heath & Jiffry, 1983; Krysinski, Ludwiczak & Mucha, 1981). Closer examination shows the importance of separating the *selection* and *breakage functions*; two tests of comminution of nuts correlated well (Manly & Braley, 1950; Helkimo, Carlsson & Helkimo, 1977) and they both correlated moderately well with the swallowable thresholds, i.e. the particle size of chewed soya beans perceived as ready to swallow (Jiffry, 1983). All of these are critically dependent upon the selection function, but contrast with one based on the extraction of sugar from chewing gum (Heath, 1972), which demands only the simpler task of kneading a viscoelastic bolus.

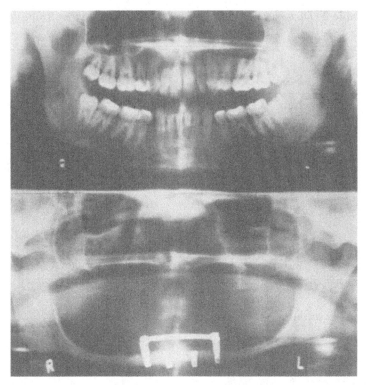

Fig. 15. Radiographs of a dentate person (top) contrasted with the profound loss of bone as well as teeth in an edentate person (bottom). The latter has titanium implants to stabilise her lower denture.

Adaptation to loss of teeth and ageing

Man's refined diet was responsible until recently for almost ubiquitous dental caries. The effects are not confined to loss of teeth because the supporting bone is then progressively lost until there is inadequate bone to stabilise dentures (Fig. 15). When surveyed in 1978, 84% of adults resident in England and Wales over the age of 74 had no natural teeth (Todd & Walker, 1980). To aggravate the difficulties, muscle wastes with age (Heron & Chown, 1967; Newton *et al.*, 1987) and the combined effects dramatically reduce masticatory effectiveness (Heath, 1982). The trend is that old people with poor unstable dentures chew poorly and slowly (Fig. 16) and select foods that are easier to chew (Table 2; and Heath, 1972).

However, with the combination of physiological reserve, adaptive

Table 2. *Dietary selection by 75 elderly denture wearers (data from Heath, 1972)*

Food	% of subjects ever eating food
Nuts	36
Tough meat	43
Hard biscuits	51
Raw celery	53
Toffee	55
Crust on bread	67
Lettuce	75
Sticky pastries	75
Oranges	75
Hard cheese	77
Minced beef	93
Fish	96

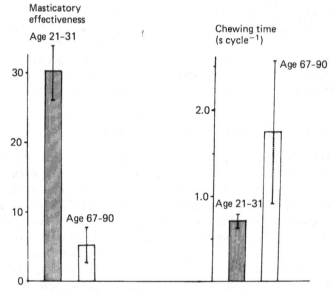

Fig. 16. Mastication in elderly denture wearers is dramatically poorer and slower than that of young dentate people. (From Heath (1982), with permission from the *International Dental Journal*.)

behaviour and good dental care, biological appetite continues to provide most of us with the pleasures of chewing.

References

Adran, G.M. & Kemp, F.H. (1955). A radiographic study of movements of the tongue in swallowing. *Dental Practitioner*, **5**, 252–61.

Ahlgren, J. (1966). Mechanism of mastication. *Acta Odontologica Scandinavica*, **24** (Supplement 44), 1–109.

Anderson, D.J., Hannam, A.G. & Matthews, B. (1970). Sensory mechanisms in mammalian teeth and their supporting structures. *Physiological Reviews*, **50**, 171–95.

Anderson, D.J. & Hector, M.P. (1987). Periodontal mechanoreceptors and parotid secretions in animals and man. *Journal of Dental Research*, **66**, 518–23.

Anderson, D.J., Hector, M.P. & Linden, R.W.A. (1985). The possible relation between mastication and parotid secretion in the rabbit. *Journal of Physiology*, **364**, 19–29.

Bates, J.F., Stafford, G.D. & Harrison, A. (1975a). Masticatory function – a review of literature. I. The form of the masticatory cycle. *Journal of Oral Rehabilitation*, **2**, 281–301.

Bates, J.F., Stafford, G.D. & Harrison, A. (1975b). Masticatory function – a review of the literature. II. Speed of movement of the mandible, rate of chewing and forces developed in chewing. *Journal of Oral Rehabilitation*, **2**, 349–61.

Bates, J.F., Stafford, G.D. & Harrison, A. (1976). Masticatory function – a review of the literature. III. Masticatory performance and efficiency. *Journal of Oral Rehabilitation*, **3**, 57–67.

Bearn, E.M. (1973). Effect of different occlusal profiles on the masticatory forces transmitted by complete dentures. *British Dental Journal*, **134**, 7–10.

Dahlberg, B. (1942). The masticatory effect. *Acta Medica Scandinavica*, **112** (Supplement 139), 1–479.

Dixon, A.D. (1963). Nerve plexuses in the oral mucosa. *Archives of Oral Biology*, **8**, 435–47.

Drake, B. (1965). On the biorheology of human mastication: an amplitude–frequency–time analysis of food crushing sounds. *Biorheology*, **3**, 21–31.

Epstein, B. (1947). The mathematical description of certain breakage mechanisms heading to the logarithmico-normal distribution. *Journal of the Franklin Institute*, **244**, 471–7.

Farrell, J.H. (1956). The effect of mastication on the digestion of food. *British Dental Journal*, **100**, 149–55.

Fish, E.W. (1934). *Principles of Full Denture Prostheses*, 2nd edn, pp. 25–32. John Bale, Sons & Danielsson, London.

Gardner, R.P. & Austin, L.G. (1962). A chemical engineering treatment of certain breakage mechanisms leading to the logarithmico-normal distribution. In *Zerkleinern Symposium*, pp. 217–48. Verlag Chemie, Dusseldorf.

Garrett, J.R. (1987). The proper role of nerves in salivary secretion: a review. *Journal of Dental Research*, **66**, 387–97.

Hand, A.R. (ed.) (1987). Editorial. In *Saliva and Salivary Glands*. Proceedings of the 10th International Conference of Oral Biology. *Journal of Dental Research*, **66**, 385–618.

Heath, M.R. (1972). Dietary selection by elderly persons, related to dental state. *British Dental Journal*, **132**, 145–8.

Heath, M.R. (1982). The effect of maximum biting force and bone loss upon masticatory function and dietary selection of the elderly. *International Dental Journal*, **32**, 345–56.

Heath, M.R. (1986). Functional interpretation of patterns of occlusal wear on acrylic teeth. *Restorative Dentistry*, **2**, 100–7.

Heath, M.R. & Bellwood, P.C. (1988). The masticatory–salivary reflex in complete denture users. *Journal of Oral Rehabilitation*, **15**, 540.

Heath, M.R. & Lucas, P.W. (1988). Oral perception of texture. In *Food Structure – Its Creation and Evaluation*, pp. 465–81. Butterworths, London.

Helkimo, E., Carlsson, G.E. & Helkimo, M. (1977). Chewing efficiency and state of dentition. *Acta Odontologica Scandinavica*, **36**, 33–41.

Helkimo, E., Heath, M.R. & Jiffry, M.T.M. (1983). Factors contributing to mastication: an investigation using four different test foods. *Journal of Oral Rehabilitation*, **10**, 431.

Heron, A. & Chown, S. (1967). *Age and Function*, p. 63. Churchill, London.

Hey, H., Jorgensen, F., Sorensen, K., Hasselbalch, H. & Wamberg, T. (1982). Oesophageal transit of six commonly used tablets and capsules. *British Medical Journal*, **85**, 1717–19.

Hsu, R. (1987). Functional masticatory forces; development and application of a hydraulic measuring device. M.Sc. dissertation, University of London.

Jemt, T. (1984). Masticatory mandibular movements. *Swedish Dental Journal*, Supplement, **24**, 35–41.

Jemt, T. & Hedegard, B. (1982). Reproducibility of chewing rhythm and of mandibular displacements during chewing. *Journal of Oral Rehabilitation*, **9**, 531–7.

Jenkins, G.N. (1978). *Physiology and Biochemistry of the Mouth*, 4th edn, pp. 532–7. Blackwell, Oxford.

Jiffry, M.T.M. (1983). Analysis of particles produced at the end of mastication in subjects with normal dentition. *Journal of Oral Rehabilitation*, **8**, 113–19.

Kerr, A.C. (1961). *The Physiological Regulation of Salivary Secretion in Man*. Pergamon, Oxford.

Krysinski, Z., Ludwiczak, T. & Mucha, J. (1981). Comparative investigations of selected methods of evaluating the masticatory ability. *Journal of Prosthetic Dentistry*, **46**, 568–71.

Laine, P. & Siirila, H.S. (1971). Oral and manual stereognosis and two-point tactile discrimination of the tongue. *Acta Odontologica Scandinavica*, **29**, 197–204.

Lightoller, G.H. (1925). Facial muscles. The modiolus and muscles surrounding the rima oris with some remarks about the panniculus adiposus. *Journal of Anatomy*, **60**, 1–85.

Lucas, P.W. & Luke, D.A. (1983*a*). Methods for analysing the breakdown of food in human mastication. *Archives of Oral Biology*, **28**, 813–19.

Lucas, P.W. & Luke, D.A. (1983*b*). Computer simulation of the breakdown of carrot particles during human mastication. *Archives of Oral Biology*, **28**, 821–6.

Lucas, P.W. & Luke, D.A. (1984). Optimum mouthful for food comminution in human mastication. *Archives of Oral Biology*, **29**, 205–10.

Lucas, P.W., Ow, R.K.K., Ritchie, G.M., Chew, C.L. & Keng, S.B. (1986). Relationship between jaw movement and food breakdown in human mastication. *Journal of Dental Research*, **65**, 400–4.

Lund, J.P. & Olsson, K.A. (1983). The importance of reflexes and their control during jaw movement. *Trends in Neuroscience*, **6**, 458–63.

Manly, R.S. & Braley, L.C. (1950). Masticatory performance and efficiency. *Journal of Dental Research*, **29**, 448–62.

Mongini, F., Tempia-Valenta, G. & Benvegnu, G. (1986). Computer-based assessment of habitual mastication. *Journal of Prosthetic Dentistry*, **55**, 638–49.

Newton, J.P., Abel, R.W., Robertson, E.M. & Yemm, R. (1987). Changes in human masseter and medial pterygoid muscles with age: a study by computed tomography. *Gerodontics*, **3**, 151–4.

Olthoff, L.W. (1986). Comminution and neuromuscular mechanisms in human mastication. Doctoral thesis, University of Utrecht.

Olthoff, L.W., van der Bilt, A., Bosman, F. & Kleisen, H.H. (1984). Distribution of particles sizes in food comminuted by human mastication. *Archives of Oral Biology*, **29**, 899–903.

Owall, B. (1978). Interocclusal perception with anaesthetized and unanaesthetized temporomandibular joints. *Swedish Dental Journal*, **2**, 199–208.

Owall, B. & Vorwerk, P. (1974). Analysis of a method for testing oral tactility during chewing. *Odontologisk Revy*, **25**, 1–10.

Pavlov, I.P. (1897). *Lectures on the Work of the Principal Digestive Glands* (transl. 1910). Griffin, London.

Porter, R. (1966). Lingual mechanoreceptors activated by muscle twitch. *Journal of Physiology*, **183**, 101–11.

Preuschoft, H. (1989). A biomechanical approach to the evolution of the facial skeleton of hominoid primates. *Fortschritte der Zoologie*, **35**, 421–31.

Ringel, R.L. & Ewanowski, S.J. (1965). Oral perception. 1. Two point discrimination. *Journal of Speech and Hearing Research*, **8**, 389–97.

Sheine, W.S. (1979). The effects of variations in molar morphology on masticatory effectiveness and digestion of cellulose in prosimian primates. Ph.D. thesis, Duke University, North Carolina.

Speirs, R.L., Staniforth, A. & Sittampalam, G. (1988). Subjective assessment of liquid volumes during swallowing. *Archives of Oral Biology*, **33**, 701–6.

Szczesniak, A.S. (1963). Classification of textural characteristics. *Journal of Food Science*, **28**, 385–9.

Taylor, A., Appenteng, K. & Morimoto, T. (1981). Proprioceptive input from the jaw muscles and its influence on lapping, chewing and posture. *Canadian Journal of Physiology and Pharmacology*, **59**, 636–44.

Todd, J.E. & Walker, A.M. (1980). *Adult Dental Health*, vol. 1 *England & Wales 1968–1978*, pp. 8–9. HMSO, London.

Tornberg, E., Fjelkner-Modig, S., Ruderus, H., Glantz, P.O., Randow, K. & Stafford, G.D. (1985). Clinically recorded masticatory patterns as related to the sensory evaluation of meat and meat products. *Journal of Food Science*, **50**, 1059–66.

Utz, K.H. (1983). The interocclusal tactile fine sensibility of natural teeth. *Journal of Oral Rehabilitation*, **10**, 440–1.

Utz, K.H. & Wegmann, U. (1985). The interocclusal tactile fine sensibility of complete denture wearers. *Journal of Oral Rehabilitation*, **12**, 549.

Voon, F.C.T., Lucas, P.W., Chew, C.L. & Luke, D.A. (1986). A simulation approach to understanding the masticatory process. *Journal of Theoretical Biology*, **119**, 251–62.

Wickwire, N.A., Gibbs, C.H., Jacobson, A.P. & Lundeen, H.C. (1981). Chewing patterns in normal children. *Angle Orthodontist*, **51**, 48–60.

Young, J.Z. (1971). *An Introduction to the Study of Man*. Oxford University Press, Oxford.

Yurkstas, A. (1951). Compensation for inadequate mastication. *British Dental Journal*, **91**, 261–2.

Yurkstas, A. (1965). The masticatory act. *Journal of Prosthetic Dentistry*, **15**, 248–60.

D. KILCAST AND A. EVES

Integrating texture and physiology – techniques

Texture is a fundamental sensory property of all foods, which is of great importance in determining consumer acceptability. The inability to formulate specific food textures is the major difficulty encountered in texture development and control. This results not only from a lack of fundamental knowledge of the factors that manifest themselves as texture, but also from a lack of understanding of the physiology of texture perception. Successful design of texture relies on having a means of defining and quantifying the many distinct textural attributes that foods possess. The perception of texture is a psychological response to a tactile stimulus, and therefore a full description of texture can be achieved only through the use of people.

The initial texture reaction is visual followed by the tactile reaction when the food is either cut or placed in the mouth (Peleg, 1980). The physiological reactions to food in the mouth can trigger a psychological reaction. The complexity is emphasised by Szczesniak & Kahn (1971), who described perception in the mouth as 'a mixture of conscious and unconscious processes, the awareness being accentuated when visual expectations are violated'.

The neurological basis of oral perception involves stimulation of at least two different sensory systems. Food presents a tactile stimulus to the tongue, palate and pharyngeal regions; and chewing, through movement of both the jaw and the tongue, is the cause of muscular sensation.

Research into texture has been beset with many difficulties, misunderstandings and conflicts, which have resulted largely from the confusion surrounding terminology, and prevented rapid progress. Kramer (1955) divided the quality factors of foods into three categories: (1) appearance; (2) kinaesthetic; and (3) flavour. Earlier, Kramer (1951) had listed chewiness, fibrousness, succulence and grittiness as the kinaesthetic factors of importance in objective testing of vegetable textural quality. Szczesniak (1963a) linked texture to sensory, structural and physical parameters and Corey (1970) said 'texture is but another name for the interaction of the human with the mechanical properties of the material'. Sherman (1970), in a modification of Szczesniak's (1963b) definition, expressed texture as 'the composite of those properties arising from the structural elements, and the manner in which they register with the physiological senses'.

Jowitt (1974) stated that the appreciation of texture involves the subtle interaction between both motor and sensory components of the masticatory and central nervous systems. In other words, the complex reactions caused during the chewing of food are all taken into account by the brain in a comprehensive description of texture.

Subjective textural measurement

The principles described by Szczesniak were incorporated into the first texture profile system to be adopted by the food industry (e.g. Szczesniak, Brandt & Friedman, 1963; Brandt, Skinner & Coleman, 1963; Civille & Szczesniak, 1973). Standard scales were devised for rating various textural parameters, such as hardness, chewiness and adhesiveness. Each scale was of the internal type and each point was identified by a reference food. Each reference food was identified by its descriptive name, brand or type, manufacturer's name, sample size and temperature of serving. Each parameter was given a careful definition. Other textural characteristics were evaluated less quantitatively. The order of appearance of the various characteristics was specified and grouped into three stages: initial (first bite), masticatory and residual. Panellists received extensive training in the terminology and methodology of profiling. Initially, they had foods that corresponded to points on the standard rating scales and later they gained experience in preparing profiles for a range of other foods.

A major practical problem facing companies using this method arises from the lengthy training procedures; for example, for training 12 panellists, about 1900 hours are required over a six month period. As a consequence, most of the industry has attempted to use profile methods that are less demanding on time and cost. The most commonly used methods are variants of the quantitative descriptive analysis (QDA) procedure, first described by Stone et al. (1974).

Unlike the previously described method, a QDA panel is normally used for the evaluation of one product type only. Potential assessors are first screened for their suitability for such work. The panel then, through discussion sessions, derives a list of descriptive terms that define the product range, using the products themselves. These terms may also include appearance and flavour terms. The panel is then trained to score the intensity of the attributes on a line scale, anchored by terms such as 'absent' and 'strong'. Training is continued until the panellists can score the attributes reproducibly and the results can be presented both numerically and graphically. Although the method is more cost effective, it is still demanding on panellists' time, and the performance of a panel needs to be maintained through ongoing training if it is used only irregularly.

An additional criticism often levelled at QDA and associated methods is that panellists are forced into a consensus on the use of terms and that as a consequence there is often confusion. Although different use of terms can often be identified by statistical methods such as principal components analysis, an alternative method, called free-choice profiling, has been described (Arnold & Williams, 1986), in which each assessor constructs his/her own set of descriptors which is then used for subsequent scoring sessions. The use of personal descriptor sets minimises the need for training. Since there is generally considerable confusion in the use of textural terms, the method has great potential for texture profiling, but suffers from the problem that the large number of individual configurations that result requires statistical techniques such as generalised procrustes analysis. The techniques are complex and are unsuitable for use on microcomputers. The practical difficulties posed by these subjective methods have resulted in the food industry's seeking simple and inexpensive objective measurements of texture.

Objective textural measurement

Objective methods of texture measurement may be divided into three main categories: fundamental, empirical and imitative (Scott-Blair, 1958).

Fundamental tests measure rheological properties, such as elastic moduli and viscosities, whereas empirical tests often measure ill-defined parameters that are indicated by practical experience to be related to some aspect of textural quality. Devices such as penetrometers, direct compression devices, consistometers and shearing devices have been used.

Imitative methods of measurement mimic the conditions to which the material is subjected in practice. Volodkevich (1938) describes the bite tenderometer, which attempts to mimic the action of teeth on food. It records the force of biting on a piece of food as a function of the deformation incurred. Two wedges with rounded points substitute for teeth, the lower being fixed to a frame. The upper wedge can be moved with a linear motion through the arc of a circle by a lever, thus squeezing a sample between the wedges.

The closest analogy to human chewing was attained by the Denture

Tenderometer (Proctor *et al.*, 1955), an adaptation of Volodkevich's apparatus utilising a pair of human dentures to enable determination of food-crushing forces. The driving mechanism was able to impart both lateral and forward motion to the lower jaw. The amplified output from strain gauges placed on the articular arm measured the forces encountered. This served as the prototype for the General Foods (GF) Texturometer (Friedman, Whitney & Szczesniak, 1963) in which the dentures are replaced by a plunger. The location of the sensing element was moved from the articular arm to the sample area to eliminate gravity forces, and the oscilloscope replaced by a chart recorder, enabling easy and permanent recording of any chosen number of consecutive chews. In this device the driving mechanism no longer imparts a combined lateral and forward motion to the lower jaw, although it still drives the plunger through the arc of a circle. The linear speed of travel of the plunger varies sinusoidally with time (Brennan, Jowitt & Williams, 1975).

Although the GF Texturometer remains in use in North America and in Japan, the Universal Testing Machines exemplified by those made by Instron and the Stevens Compression Response Analyser are most commonly used in the food industry. The instruments differ in their mechanical construction and in their data acquisition and data analysis capabilities, but they have a number of important features in common. All have a cross-head containing a load cell, which is driven vertically at a range of constant speeds, and which can cycle over a fixed distance or load range. Probes can be attached to the cross-head for penetration, shearing or crushing food, which can be held in a variety of cells. The load is recorded relative to time or to penetration/deformation distance, and displayed on a suitable recorder. Computer control of the instrument and computer analysis of the data are increasingly common.

Two major difficulties are associated with the use of such instruments. Firstly, they are too often used uncritically, without any attempt being made to ensure that the parameters used are relevant to the sensory perception of texture. Secondly, the instrument-operating conditions bear little relationship to conditions in the mouth during chewing, and textural characteristics associated with geometrical factors and breakdown factors are not measured at all.

Physiological aspects of chewing

Once food is bitten with the incisors or placed in the mouth, a complex series of changes occurs. Firstly, the food may be positioned in an appropriate part of the mouth using tongue movement. Pressure from the tongue may press the food against the hard palate, giving information on the response of the food to such pressure and on the surface textural characteristics. Biting using the lateral sets of teeth occurs in two stages, the first of which occurs at low loads with a high rate of movement (up to 400 cm min^{-1} for whole meat; Tornberg *et al.*, 1985) and the second at high loads with low speeds (Brennan, 1988). The motion itself can be composed of complex combinations of vertical and horizontal movements, and can depend on the state of occlusion of the teeth (Boyar & Kilcast, 1986). These movements lead to the food being subjected to a complex combination of compressive, shearing and tensile forces.

During this process, saliva is being introduced to the food, acting as a lubricant and helping in the formation of a bolus with the correct physical properties needed for swallowing (Szczesniak, 1987). Foods initially at temperatures higher or lower than body temperature also undergo physical changes when eaten as thermal equilibration occurs. The fatty component of meat products, for example, can solidify, which gives a change to the mouthfeel characteristics. More importantly, chilled and frozen foods undergo changes leading to melting of any ice component and extensive structural breakdown; this can happen rapidly even in the absence of chewing.

It is clear that no mechanical instrument can simulate this wide range of conditions. At best, an instrument can simulate only a first bite, and even then only with an oversimplified mechanical action. Moreover, few instruments can reach the high force application rates experienced in the mouth; most tests in the industry are operated at 3 cm min^{-1}. Since most solid foods are viscoelastic in nature, rate of force application is an important factor in measuring physical properties, but is rarely taken into account.

An elegant solution to the problems surrounding sensory assessments using people, and objective measurements using instruments, would be to make objective measurements on people during eating. One approach has been to analyse the sound emitted when certain foods are bitten and chewed (e.g. Drake, 1963; Vickers, 1985). Such studies have indicated that vibratory stimuli could form the basis for crispness determination, and that crisp foods could be distinguished from crunchy foods, crisp foods giving sounds that are both louder and higher in pitch.

Such an approach is, of course, limited both to foods that emit sounds during chewing and also to a small number of textural attributes. Electromyography (EMG) has been used for some time for studying muscular activity, but little work has been carried out to use the method for determining the activity of muscles used during chewing and relating the activity to food texture. The remainder of this chapter describes some of the work carried out in this important area at the Leatherhead Food Research Association.

Methods for recording electromyographic data

Electromyographic patterns have been recorded using a Grass Polygraph (Model 7D). The system has a regulated power supply, consisting of two DC driver amplifiers. One of these is connected to a pre-amplifier and displays the raw signal. The other is connected to a pre-amplifier and an integrator module and displays the integrated data. Both driver amplifiers include a 50 Hz filter rejecting interference from AC sources. All recordings were made with the amplifiers' frequency bands as wide as possible (10–40 kHz). This prevents the recording from undulating with the movement of the jaw and ensures a flat frequency response up to 200 Hz, this being the maximum response rate of the pens. The time base was set at 0.2 s at the recommendation of the manufacturer.

Surface electrodes were used to detect the electrical signals from the main muscle used in chewing solid foods – the masseter muscle. This muscle is shallow-lying and activity can be recorded readily without the need for implanted needle electrodes.

The skin on the cheek and earlobe were cleansed with 95% (v/v) ethanol to remove traces of dirt and perspiration, which can interfere with the signal. A position was located on the cheek at the maximum point of inflection of the masseter muscle and two electrodes were located above and below this point in line with the muscle and approximately 0.5 cm apart. A third electrode was placed on the earlobe, a point of no muscular activity, which acted as an earth. Electrode cream, a conductive paste, was applied to the electrode surface in all cases.

After the terminal ends of the electrodes had been located in the Polygraph, the subject was presented with the sample to chew. The data acquisition system was capable of recording information from up to three muscles simultaneously; however, the experiments described in this chapter were carried out using a single masseter muscle, and consequently the subjects were asked to eat on the side of that muscle only. All samples were assessed in triplicate and all samples were of the same size and geometry. Data were recorded from the time the sample was put into the mouth to the time of swallowing.

The raw EMG signal and its integrated form were recorded on a Y-t recorder, together with a time signal (Fig. 1). The integrated wave was digitised and stored on an IBM PCAT microcomputer using a commercial data acquisition package, ASYST (Keithley Instruments, Reading),

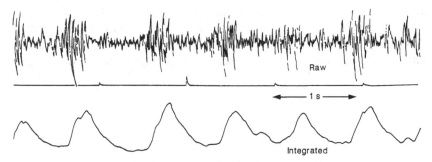

Fig. 1. EMG output.

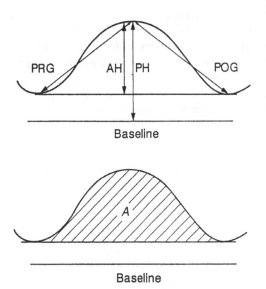

Fig. 2. Measured EMG parameters on the integrated signal. PH, peak height; AH, adjusted height; PRG, pre-maximum gradient; POG, post-maximum gradient; A, area.

modified for this purpose. A digitisation rate of 50 points per second was used, and from the data a number of curve parameters were calculated – peak height, pre-maximum gradient, post-maximum gradient and area under peak (A) (Fig. 2).

Gradients and area were calculated from minima rather than baseline and peak heights were adjusted in relation to a prerecorded baseline. Any of the trace parameters can be plotted against time, and the package permits the fitting of a range of curves to the data. With the exception of the post maximum gradient, however, the change in most EMG parameters with time can be described by either a falling exponential curve or by a rising straight line followed by a falling exponential curve. The following broken-stick model was consequently fitted to peak height data:

$$y = \{Ae^{-\alpha u}+Q(t-u)\} \text{ if } t<U$$
$$y = Ae^{-\alpha t} \text{ if } t \geq U$$

Interpretation of integrated peak height

In an experiment designed to investigate the physical significance of integrated peak height, a strain gauge force transducer was constructed and calibrated under compression using a Stevens Compression Response Analyser. A subject was linked to the Polygraph and asked to bite on the tip of the strain gauge using the molars. The voltage produced from the strain gauge was recorded, together with the integrated peak height. A plot of biting force against peak height is shown in Fig. 3. A good straight line relationship is seen over most of the range tested ($r = 0.96$), indicat-

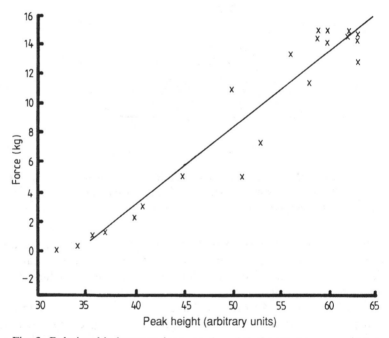

Fig. 3. Relationship between integrated peak height and measured bite force.

ing that the peak height gives a good measure of biting force in the mouth.

Between-subject reproducibility

Thirty subjects (10 males, 20 females) were asked to chew samples of fruit pastilles whilst linked to the Polygraph. Each subject chewed a total of three replicate samples, and integrated peak height data were averaged and plotted against time. Results showed that 26 subjects gave data that could be described by a broken-stick relationship (Group 1), although subjects within this group exhibited a wide range of amplitudes and chewing times. The other four subjects fell into two groups – those who gave data best represented by a single exponential decay curve (Group 2) and those who produced increasing forces during chewing (Group 3). Illustrative examples of the three curve types are shown in Fig. 4.

Experience of the technique has shown that the choice of subjects for EMG work is not critical provided they are chosen from Group 1. The

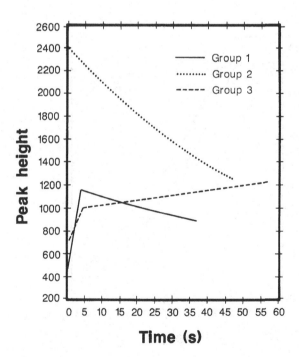

Fig. 4. Integrated peak height – time plots showing behaviour of 26 subjects.

behaviour characteristic of Group 2 may be a result of aggressive biting behaviour – the upward phase is thought to be a consequence of either positioning the food before biting proper, or a tentative biting of foods that are unfamiliar or that are recognised as hard or chewy. Research has concentrated predominantly on the breakdown phase, but valuable information on oral manipulation may be contained in the upward phase. The Group 3 behaviour, which implies increased biting forces as food is broken down, appears to be anomalous and is worthy of further investigation.

Between-occasion reproducibility

Five subjects were asked to eat 4 g samples of toffee (three replicates) on three separate occasions. Integrated peak height was measured for each subject for each day and a broken-stick model fitted.

An example of the results obtained is shown in Fig. 5. The results apply to one subject only; similar results were, however, obtained for other

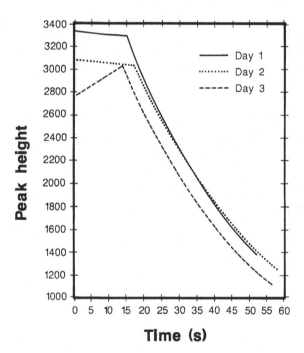

Fig. 5. Integrated peak height – time plots showing reproducibility when measured on different occasions.

subjects. Although differences were apparent in the initial phases of chewing, the breakdown rates of the samples over the three days remained the same. A slight shift is seen in the amplitude of the plot on day 3; this would probably be due to slightly inaccurate electrode placement. Chewing time on day 1 was slightly shorter than on the following two days.

From these results it would seem that from day to day results do not vary considerably. Methods by which the amplitude of peaks can be maintained are under consideration, for instance the use of a standard material on which a subject would chew prior to any study.

Further analysis of EMG data

Any non-subjective method of food texture assessment can be of practical value to the food industry only if the results of the data measured can be interpreted in subjective terms. Although the use of integrated peak height as a measure of biting force is of great value, it is clear that a considerable amount of additional textural information is contained within the EMG traces. A preliminary study was carried out in an attempt to relate the texture profile of three types of commercial confectionery gum to measured EMG parameters.

Profile analysis was carried out on three commercial fruit pastilles using the QDA method. A panel of 12 people derived textural terms for the pastilles (jelly, tough, rubbery, chewy, moist and sticky) and scored their intensities in five replicate sessions. Eleven of the 12 panellists then chewed the pastilles in three replicate sessions whilst wired to the EMG. Initial attempts were made to find simple correlations between profile terms and EMG parameters. It was found, however, that each individual profile term showed high correlations with several EMG terms and that these correlations could not be explained logically. Similarly, attempts to relate each profile term to linear combinations of EMG parameters using multiple linear regression required, typically, more than five EMG parameters to give a satisfactory fit. Such ambiguous relationships are of little practical value to the food industry, and consequently the EMG data were examined further using a multivariate statistical technique, canonical variates analysis (CVA).

Multivariate methods seek to examine patterns in data points by finding linear combinations of the original experimental variables that account for most of the variations in the data. CVA operates in a similar way, but defines groupings in the data by maximising the ratio of the variation between groups to the variation within groups. The first canonical variate (CV1) is the direction that maximises this ratio; the second

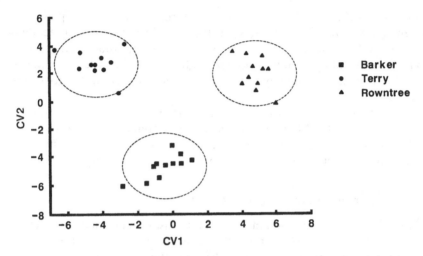

Fig. 6. Canonical variate 1 (CV1) versus canonical variate 2 (CV2) for EMG data measured on commercial gums.

(CV2) is the orthogonal direction that maximises the remaining ratio, and so on.

Figure 6 shows the CVA plot from analysis of EMG parameters for the three commercial fruit pastilles. Each point represents one subject, and the dotted lines represent the 95% confidence limits for each grouping. The plot separates the three pastilles into three non-overlapping groups with most subjects falling within the 95% confidence limits. Since CVA is designed to maximise the separation between groups, a check was carried out on the risk that the groupings might be an artefact of the method. Nine identical samples of wine gums were chewed and the EMG data treated as three different samples chewed in triplicate. The resultant CVA plots are shown in Fig. 7. The overlap of the 95% confidence circles shows that no distinct groupings were found in the data.

In a second set of experiments designed to examine further the use of the CVA method, profile and EMG analyses were carried out on a set of confectionery materials formulated and manufactured to give a controlled range of structures and textures. The set comprised two wine gum formulations (50/50 gelatin/starch and 100% gelatin, respectively), a gelatin chew and a whipped chew. The QDA profile generated a total of 15 textural terms. EMG data for peak height are shown in Fig. 8, and CVA plots on all EMG parameters are shown in Fig. 9. The CVA plots show the gelatin/starch wine gum to be well separated from the other three products, all of which overlap. The reason for this separation may

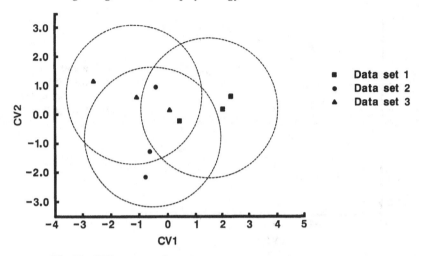

Fig. 7. CV1 versus CV2 for EMG data from nine replicates of one sample. Abbreviations as for Fig. 6.

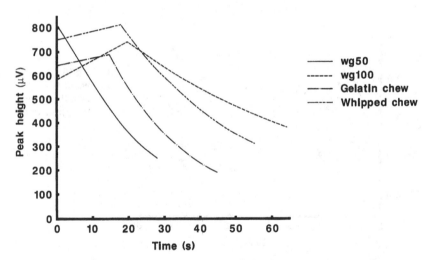

Fig. 8. Peak height data from four confectionery products with different structural and textural characteristics. wg50 – gelatin/starch wine gum; wg100 – gelatin wine gum.

be associated with the unique breakdown characteristics seen in Fig. 8, in which the gelatin/starch wine gum exhibits rapid breakdown with no initial upward phase. CVA plots on the profile data are shown in Fig. 10.

This analysis separates the gelatin/starch wine gum from the gelatin and

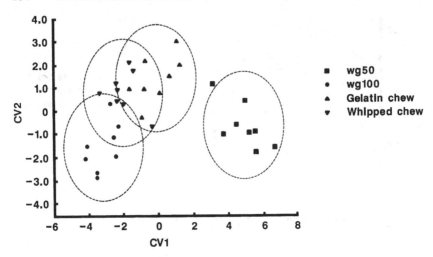

Fig. 9. CV1 versus CV2 for EMG data from four confectionery products. wg50 and wg100, see Fig. 8; CV1 and CV2, see Fig. 6.

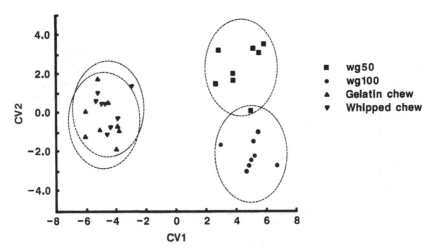

Fig. 10. CV1 versus CV2 for profile data from four confectionery products. wg50 and wg100, see Fig. 8; CV1 and CV2, see Fig. 6.

whipped chews, but not from the gelatin wine gum sample. One reason for the difference between the CVA plots on the EMG and profile data may be that the profile data contain textural information on attributes, such as grainy and floury, that may not influence the EMG data.

Although considerable effort remains to be made in understanding the potential and the limitations of EMG, the use of CVA enables foods of different textural characteristics to be readily visualised in a low dimensionality texture space. This gives the product developer a powerful tool in attempting to match the textural characteristics of foods and in identifying novel textures not represented in CVA texture space.

Summary

The understanding of food texture and its importance to the consumer is considerably less well advanced than the understanding of factors such as odour and taste. Previous attempts to develop an improved understanding have been based on physiological principles, but have not until recently developed further than an empirical approach. More extensive knowledge of the principles and methods of human physiology has led to new paths in texture research, with the emergence of the use of the human subject as a unique measuring instrument. Further exploitation of physiological principles and methods will undoubtedly lead to other innovations, which, like EMG, will be of practical value to the food industry, but such progress will require close cooperation between food scientists, physiologists and researchers from other disciplines.

Acknowledgment

The authors thank the Ministry of Agriculture, Fisheries and Food for providing financial support for this study.

References

Arnold, G.M. & Williams, A.A. (1986). The use of generalised procrustes techniques in sensory analysis. In *Statistical Procedures in Food Research*, ed. J.R. Piggott, pp. 233–53. Elsevier Applied Science, London.

Boyar, M.M. & Kilcast, D. (1986). Food texture and dental science. *Journal of Texture Studies*, **17**, 221–52.

Brandt, M.A., Skinner, E.Z. & Coleman, T.A. (1963). Texture profile method. *Journal of Food Science*, **28**, 404–9.

Brennan, J.G. (1988). Texture perception and measurement. In *Sensory Analysis of Foods*, 2nd edn, ed. J.R. Piggott, pp. 69–102. Elsevier Applied Science, London.

Brennan, J.G., Jowitt, R. & Williams, A. (1975). An analysis of the action of the General Foods Texturometer. *Journal of Texture Studies*, **6**, 83–100.

Civille, G.V. & Szczesniak, A.G. (1973). Guidelines to training a texture profile panel. *Journal of Texture Studies*, **4**, 204–23.

Corey, H. (1970). Texture in foodstuffs. *CRC Critical Reviews of Food Technology*, **1**, 161–98.

Drake, B. K. (1963). Food crushing sounds. An introductory study. *Journal of Food Science*, **28**, 233–41.

Friedman, H. H., Whitney, H. & Szczesniak, A. S. (1963). The Texturometer – a new instrument for objective texture measurement. *Journal of Food Science*, **28**, 390–403.

Jowitt, R. (1974). Terminology of food texture. *Journal of Texture Studies*, **5**, 351–8.

Kramer, A. (1951). Objective testing of vegetable quality. *Food Technology*, **5**, 265–9.

Kramer, A. (1955). Food quality and quality control. In *Handbook of Food and Agriculture*, p. 733. Reinhold Publishing, New York.

Peleg, M. (1980). A note on the sensitivity of fingers, tongue and jaws as mechanical testing instruments. *Journal of Texture Studies*, **10**, 245–51.

Proctor, B. E., Davison, S., Malecki, G. J. & Welch, M. (1955). A recording strain gauge tenderometer for foods. I. Instrument evaluation and initial tests. *Food Technology*, **9**, 471–7.

Scott-Blair, G. W. (1958). Rheology in food research. *Advances in Food Research*, **8**, 1–56.

Sherman, P. (1970). The correlation of rheological and sensory assessment of consistency. In *Industrial Rheology*, pp. 371–91. Academic Press, London.

Stone, H., Sidel, J., Oliver, S., Woolsey, A. & Singleton, R. C. (1974). Sensory evaluation by quantitative descriptive analysis. *Food Technology*, **28**(11), 24–34.

Szczesniak, A. S. (1963a). Objective measurement of food texture. *Journal of Food Science*, **28**, 410–20.

Szczesniak, A. S. (1963b). Classification of textural characteristics. *Journal of Food Science*, **28**, 385–9.

Szczesniak, A. S. (1987). Relationship of texture to food acceptance and nutrition. In *Food Acceptance and Nutrition*, pp. 157–72. Academic Press, London.

Szczesniak, A. S., Brandt, M. A. & Friedman, H. H. (1963). Development of standard rating scales for mechanical parameters of texture and correlation between the objective and sensory methods for texture evaluation. *Journal of Food Science*, **28**, 397–403.

Szczesniak, A. S. & Kahn, E. L. (1971). Consumer awareness and attitudes to food texture. I. Adults. *Journal of Texture Studies*, **2**, 280–95.

Tornberg, E., Fjelkner-Modig, S., Ruderus, H., Glantz, P., Randow, K. & Stafford, D. (1985). Clinically recorded masticatory patterns as related to the sensory evaluation of meat and meat products. *Journal of Food Science*, **50**, 1059–66.

Vickers, Z.M. (1985). The relationships of pitch, loudness and eating technique to judgements of the crispness and crunchiness of food sounds. *Journal of Texture Studies*, **16**, 85–95.

Volodkevich, N.N. (1938). Apparatus for measurement of chewing resistance or tenderness of foodstuffs. *Food Research*, **16**, 73–82.

A. C. SMITH

Brittle textures in processed food

Processes and products

Brittle products are associated with a wide range of raw materials. In the confectionery area chocolate, toffee and boiled sweets are products which may break at low strains under certain conditions of composition, temperature and strain rate. Brittle potato products include crisps directly produced from the tuber and various snacks produced from the intermediate of dehydrated potato, although cereals offer probably the greatest variety of brittle products from a number of processing operations.

Brittle products are taken here to include the textural terms 'crisp', 'hard' and 'crunchy'. An alternative definition of 'brittle' may be found in the polymer engineering literature. Powell (1983) characterised brittle failure by the creation of new surfaces and fragmentation after the polymer has deformed to a small extent and Ward (1983) designated brittle behaviour from the stress–strain response as failure at the maximum load which would normally occur at low strains. In contrast ductile behaviour results in a load maximum before failure. Rapid loading tends to favour brittle failure; the implication of the rate of eating in the texture of foods is a further complication.

Composites

Different textures may be produced by the fabrication of composites. Foams represent a special case where the material occupies the foam walls; the lower density of a foam results in reduced stiffness and strength compared to an isotropic solid with correspondingly modified textural properties (see Jeronimidis, this volume). In general, composites make use of components of different properties and it is interesting to note that in synthetic materials engineering inclusions are made in the matrix to toughen or stiffen the component. Orientation in the matrix or the filler may be used in the construction of anisotropic textural properties. Cakes have been considered as composites comprising starch grains in a protein matrix. In foods, brittle failure may be designed to occur selectively; for example, chocolate sheets in ice cream show a brittle response which confers a unique texture to the product. Coextrusion may be used to form brittle tubes filled with a soft cream having markedly different overall properties compared to a hollow tube, which raises the question of equilibrium between adjacent different materials. A more subtle composite is that of a product which has an uneven moisture distribution resulting in a hard exterior and soft interior. Multi-textured products may then be designed with moisture barriers to maintain a non-equilibrium distribution.

The design of brittle products must also include the geometry of the product, the inclusion of stress concentrators such as holes and sharp corners reducing the force necessary for failure to occur. More obviously the scale and shape of the product will affect the failure stress (Powell, 1983); the role of flaws in controlling fracture is discussed in detail elsewhere (Ward, 1983).

Operations

Some of the principal processes for producing products which are brittle at ambient temperature are described below. The production of a food having brittle mechanical properties may involve a combination of operations: for example, multiple drying steps or a sequence of forming, drying and frying.

Flaking

Flaking of cereals may be carried out on whole grains, large particles or flours. If grains are steamed they can be deformed into thin plates or flakes by rolling. Cooking of moist flours followed by compaction into pellets may also precede rolling to produce flakes and then drying or toasting to produce the final product (Fig. 1).

Pressure vessels may be used to subject cereal grains to a pressure and temperature history to achieve cooking. Breakfast cereals are often prepared in this way for subsequent processing by rolling, cutting and toasting. The initial stage of the process may, however, be carried out by continuously forming the material following a high temperature–short time process (Fig. 2).

Baking

Baking is one of the oldest processing routes; biscuits are a brittle product produced by baking. Bread and cakes use a high protein flour and have

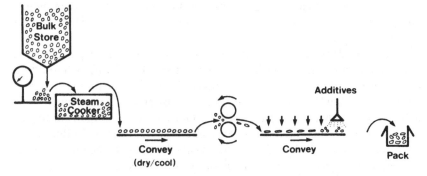

Fig. 1. Schematic illustration of flaking line.

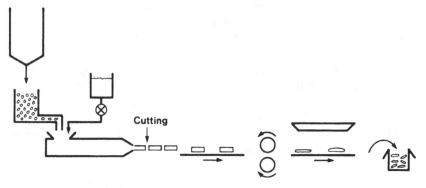

Fig. 2. Schematic illustration of flaking line using continuous extrusion cooking.

'soft' textures which may none the less become harder if not brittle with time, with staling, drying and toasting generally increasing strength and stiffness.

Baking may be carried out through heat transfer by conduction, convection and radiation and is often achieved commercially using tunnel ovens to effect continuous operation. The drying of the surface, however, retards the baking process. Direct heating methods such as radio and microwave frequency heating generate heat within the product and a combination of direct and indirect heating has certain advantages. Conventionally fired ovens are used during the early stages of baking when moisture levels are high and the material is a good conductor.

Forming and moulding
Forming and moulding operations are central to the production of many foods where the filling or forming step is preceded by a combination or

mixing stage in which the ingredients are brought together. Heat generation and direct heating may be used to cook the material, which is often subjected to pressure.

Using cereals, starches or dehydrated potato a dough or paste may be formed which is partitioned to the product size by cutting or moulding. For example, wafers are formed from flour and water mixtures heated between plates where pressure is generated by steam. Many snack products are based on a dough which is rolled into a sheet and then cut to size, or alternatively the dough is prepared and formed into the product shape by pumping through dies. This is the basis of extrusion techniques which may be used to mix the ingredients and subject them to a pressure and temperature history to yield the material in a form to be shaped (Richmond & Smith, 1987). Subsequent processing may be used to produce a brittle product such as drying, toasting or frying.

Boiled sweets are produced under vacuum by boiling sugar solutions, which extracts the water to leave a solids content of approximately 98% (Lindley, 1988). Continuous processing by extrusion cooking has resulted in a number of confectionery products for which only the amount of moisture desired in the final product is added. For example, Van Zuilichem et al. (1985) produced confectionery from sucrose and liquid glucose mixtures. Chocolate is formed in moulds and in surface coatings using enrobing (coating) technology.

Expansion
Many low-density, foamed, brittle foods are produced by puffing processes which involve creating superheated water under pressure in the food matrix followed by its release as steam on reducing the pressure. This causes a significant volume expansion. Batch puffing vessels have led to continuous devices which control the residence time at the elevated pressure. Extrusion cooking also provides a means of producing foamed brittle foods where, in addition to heating the material and developing a pressure, mechanical energy is imparted to achieve mixing and controlled material degradation. Microwaves and frying may be used to expand products such as snack collets. In both cases, the temperature of the material is raised rapidly so that the water is converted to steam and expands the softened matrix.

Drying
Drying is one of the most important techniques for food preservation and also encompasses a wide range of materials and types of equipment. Drying is basically a separation process, carried out using a number of

techniques. These include (1) the removal of water by mechanical expression, (2) evaporation at atmospheric or reduced pressures, and (3) freezing water in a material and subliming the ice. Drying is often the final step in a process comprising a number of operations and interfaces most closely to the consumer.

Materials

Having outlined some of the processing operations available to fabricate brittle products, some basic materials are now examined together with their functional properties, beginning with synthetic polymers from which much can be learned about the origins of brittle properties.

Polymer properties and molecular relaxations

Polymers are classically described as amorphous or semi-crystalline. A polymer having amorphous and crystalline components shows a glass transition T_g and a crystalline melting transition T_m temperature. The glass transition increases markedly with increasing low degrees of polymerisation but is independent of molecular weight at high values. Solvents acting as plasticisers enhance molecular mobility of polymeric segments and reduce both T_g and T_m. This is exploited in solvent-induced crystallisation where T_g is lowered such that $T_g < T < T_m$ and molecular rearrangement into crystals may occur. In thermally induced crystallisation, where the temperature is raised or lowered such that $T_g < T < T_m$, temperature and time may be manipulated to obtain different degrees of crystallinity and mechanical properties. Figure 3 shows an example of polyethylene terephthalate (PET), crystallised from the melt (Groeninckx, Berghmans & Smets, 1976). Polymers have a number of relaxations associated with different bond and molecular units. Over a wide range of temperature or frequency, polymeric materials exhibit more than one relaxation region and it is often possible to postulate the nature and location of the responsible motional groups. The molecular motions which give rise to secondary relaxations are intimately related to transitions in bulk deformation behaviour. The magnitude and temperature of the relaxations depend on polymeric structure, preparation route and history. Hiltner & Baer (1974) reported studies on model systems with a structural resemblance to proteins and showed the influence of water on the relaxation peaks in a glutamic acid–leucine copolymer.

The connection between molecular relaxations and mechanical

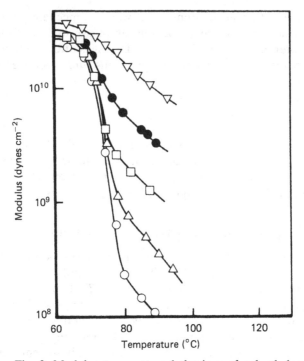

Fig. 3. Modulus–temperature behaviour of polyethylene terephthalate crystallised from the melt at 227°C as a function of time: (▽) 30 min, 50% crystallinity; (●) 27 min, 40%; (□) 22 min, 28.5%; (△) 11 min, 14.5%; (○) amorphous. (From Groeninckx *et al.*, 1976, with permission.)

properties has been shown on a number of occasions in synthetic polymer systems. The flexural modulus of PET falls dramatically at a temperature within a few degrees of the glass transition (Groeninckx *et al.*, 1976; and see Fig. 3). Polycarbonate has α (glass), β and γ transitions at 160, 80 and −110 °C, respectively (Golden, Hammant & Hazell, 1967) and annealing treatments below the glass transition increase the mechanical properties. The occurrence of brittle–ductile transitions has been discussed in detail by Ward (1983). On the assumption that brittle fracture and plastic flow are independent processes, whichever occurs at the lower stress will be operative (Fig. 4). The brittle failure mode is much less affected by strain rate and temperature than yielding but the fracture stress increases with decreasing flaw size. The effect of strain rate on the brittle–ductile transition is illustrated in Fig. 4. The relationship between the brittle–ductile

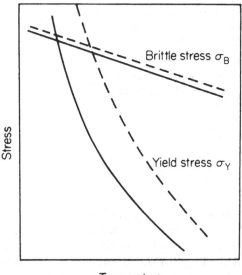

Temperature

Fig. 4. Schematic diagram of fracture stress, σ_B, and yield stress, σ_Y, as a function of temperature and strain rate. The brittle–ductile transition occurs at the intersection of the σ_B and σ_Y curves. Dashed line indicates higher strain rate than continuous line. (From Brown & Ward, 1983, with permission.)

transition and the dynamic molecular behaviour is not particularly well defined as a result of the different strains associated with their measurement. Two brittle–ductile transitions were observed in linear polyethylene by Brown & Ward (1983).

The influence of pressure has been studied in detail by Jones Parry & Tabor (1973). The application of a hydrostatic pressure hinders displacements and rotations, which means that relaxations and particularly the glass and melting transitions are shifted to higher temperatures.

The molecular transitions of food materials have received less extensive study, although recently the glass transition has received a degree of prominence in the published literature. It has been measured for specific materials as outlined below and its relevance in processing has been emphasised by Levine & Slade (1988) (some examples are listed in Table 1). They claim that this approach may be used to predict technological performance, product quality and stability for polymeric foods (Slade & Levine, 1988).

Table 1. *Some processing phenomena
related to the glass transition (from
Levine & Slade, 1986)*

Sticking, caking of amorphous powders
Structural collapse in freeze-dried products
Stickiness in spray and drum-drying
Graining in boiled sweets

Characteristics of specific materials

Starch

Starch is the principal component of cereals and vegetables and is a major
constituent of foods. It is extracted from the different indigenous sources
and used as an ingredient in its own right. Starch occurs as granules which
vary in size and shape according to their origin and contain amylose, a
linear polymer, and amylopectin, a branched polymer. Starch granules
absorb little water but on heating in an excess of water the granules swell
and absorb water, the amylose leaches out of the granule and the col-
lapsed granules containing amylopectin are held in an amylose matrix.
This process is called gelatinisation and may be detected by a number of
techniques including differential scanning calorimetry (DSC): the
gelatinisation temperatures are typically in the range 60–80°C. On reducing the water content a second endotherm is seen by DSC (Donovan,
1979), which is attributed to melting of the starch crystallites and shifts to
higher temperatures with decreasing moisture content. Starch is a
partially crystalline polymer and while the first-order melting transition
has been documented, less attention has been given to the behaviour of
the amorphous part of starch. Zeleznak & Hoseney (1987) documented
data for the T_g of native and pregelatinised wheat starches which
increased with decreasing moisture content and increasing degree of
crystallinity.

The situation under processing conditions is much less clear. The effect
of pressure on starch behaviour has received negligible attention,
although Muhr, Wetton & Blanshard (1982) found that the gelatinisation
temperature of wheat and potato starches increased by a few degrees and
then decreased on application of a pressure of 400 MPa. In addition the
quantitative effect of shear on orientation and degradation is unknown;
the starch granules may be swollen, gelatinised, disrupted or the constituents degraded.

The amylose fraction of starch is able to form complexes with lipids which increase the resistance to degradation and may be identified, for example by X-ray analysis. However, the combination of variables in the extrusion cooker has led to a new X-ray pattern which typifies a modified amylose–lipid complex (Mercier *et al.*, 1979).

Proteins

Proteins are one of the most complex of food components and are bio-polymers of amino acids which are linked through peptide bonds. Commercially important sources of vegetable proteins are soya and gluten; gluten is a complex of two protein groups, the glutenins and the gliadins. DSC reveals a glass transition temperature which falls with increasing moisture content (Hoseney, Zeleznak & Lai, 1986).

Proteins may be categorised into: (1) globular, temperature-sensitive proteins such as whey and soya bean; and (2) random-coil, temperature-insensitive proteins such as casein (Visser, 1988). The first type can be texturised by heating and the second by spinning methods. Proteins may be textured by a number of processes and these have been reviewed by Lillford (1986).

Although heating is not necessary in the formation of network structures, alternative conformations become available at the denaturation temperature. Denaturation may be termed a destruction of the native structure in which cross-links are disrupted and the protein molecules unfold; orientation may cause association with neighbouring molecules to form a network.

Sugars

Sucrose is the most abundant free sugar in plants; it may be used as a dissolved syrup or as a particulate solid and is broken down into fructose and glucose by acid treatment, termed inversion. The rate of sucrose solubility varies with temperature, degree of undersaturation and crystal size; sucrose solutions supersaturate readily and crystal growth is accelerated by deformation.

Water

Ice features in the embrittlement of many foods. Water binds strongly to polar groups of food components, which occurs in competition with water–water binding. It is highly volatile under many food processing conditions and its presence in the atmosphere leads to exchange of water which particularly affects storage. The general effect of water on the glass transition, with reference to starch, is illustrated in Fig. 5.

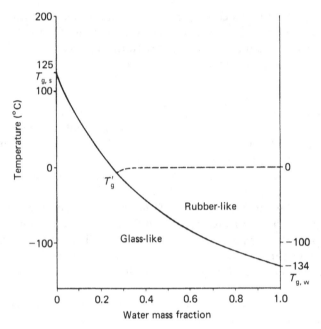

Fig. 5. The approximate glass–rubber transition temperatures as a function of mass fraction for the starch–water system. (From Van den Berg, 1986, with permission.)

Colligative and thermodynamic properties

When a solid dissolves in a liquid, the boiling point is raised and the osmotic pressure, vapour pressure and freezing point are lowered. The vapour pressure is important in defining the water activity, a_w, of foods:

$$a_w = P/P_o$$

where P is the partial pressure of water in the food at a given temperature and P_o is the vapour pressure of water at the same temperature. The relationship between a_w and moisture content, called a sorption isotherm (Fig. 6) is useful for indicating critical moisture content values (Gal, 1983) such as the processing end point, the storage limit and that corresponding to equilibrium with the environment. Slade & Levine (1988) have recently described the use of glass transition and related concepts to give information on the state of the water. Non-equilibrium situations, for example starch gelatinisation, protein denaturation and sugar crystallisation, impose a serious limitation on a thermodynamic interpretation of sorption isotherms. Most water sorption data have been obtained at

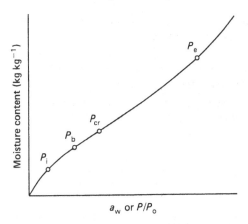

Fig. 6. Schematic sorption isotherm with typical marking points: P_i, initial point of the packaged food or end point of processing; P_b, end of the first curved part of the isotherm; P_{cr}, critical point not to be exceeded during storage and distribution; P_e, point corresponding to equilibrium with the environmental atmosphere; a_w, water activity; P, partial pressure of water in food; P_o, vapour pressure of water at same temperature. (From Gal, 1983, with permission.)

temperatures close to ambient and hence there are few data at low and high temperatures appropriate to processing. A further complication is that most sorption information relates to the hydration of dried material whereas in many processes it is desorption of water that is important and this is generally not equivalent to absorption through hysteresis effects.

Changes in phase are a frequent feature of food processing operations. The partial properties of individual components need to be known and this is often one of the most difficult and least accurate tasks in process design. In planning the processing of a food material the behaviour of the system is required in terms of the nature and composition of the phases which appear when cooling, heating or separation occur. This information dictates the composition of the food after such processing operations as freezing, drying and freeze-drying. Table 2 lists the type of thermodynamic data associated with operations involving the removal of water.

For a closed system, the condition that the specific Gibbs functions of phases at equilibrium should be equal leads to the Clausius–Clapeyron equation (Atkins, 1968). This gives the rate at which pressure must change with temperature for two phases to be in equilibrium. The case of a general system of p phases and C components leads to the definition of the chemical potential, g_i, for each constituent of each phase and the condition for equilibrium:

Table 2. *Thermodynamic and transport properties required for two water removal operations*

Drying	Enthalpy, latent heat, specific heat, volumetric properties, thermal conductivity, thermal diffusivity, sorption
Evaporation	Phase diagram, specific heat, latent heat, enthalpy, vapour pressure, volumetric properties

$$g_i^1 = g_i^2 = g_i^3 = g_i^P$$
$$\text{or } \mu_i^1 = \mu_i^2 = \mu_i^3 = \mu_i^P \tag{1}$$

where μ_i is the molar partial potential.

The chemical potential is analogous to the specific Gibbs function of a pure substance and in a phase containing a single component they are identical. However, for a mixture the chemical potential depends on concentration and the nature and proportions of the other components. This approach to drying of cellular materials is illustrated in a subsequent section.

Rheological properties

The rheology of concentrated food materials has been described by extensions of basic viscosity–strain rate relationships. A common starting point is the power law relationship:

$$\eta = k\gamma^{n-1} \tag{2}$$

where μ is the shear viscosity, γ is the shear rate, and n and k are the power law and consistency indices, respectively. In processing, the situation is made more complex when the shear rate takes different values at different points in the process. In the example of a screw pump feeding to a forming die, a shear rate of $\pi DN/h$ is associated with the flow in the screw channel (where N is the screw speed, D is the screw diameter and h is the channel depth) and a shear rate of $4Q/\pi R^3$ with flow in the die (where Q is the volumetric flow rate and R is the die radius). The shear viscosity will therefore be changed by screw speed and flow rate (Senouci & Smith, 1988a). The temperature, T, effect is often introduced through an Arrhenius exponential term to give:

$$\eta = K \exp E/RT \tag{3}$$

A variation of equation (3) which includes reference to T_g is (Bin Ahmad & Ashby, 1988):

$$\eta = K \exp \left[\frac{E}{RT_g} \left(\frac{T_g}{T} - \frac{T_g}{T_o} \right) \right] \tag{4}$$

where K is the viscosity at a reference temperature T_o.

The Williams–Landel–Ferry (WLF) equation has been used to provide a description of the viscosity in the region $T_g < T < T_m$:

$$\eta = K \exp \left(\frac{A(T-T_o)}{B+T-T_o} \right) \tag{5}$$

where A and B are constants. The reference temperature T_o is often equated to T_g.

Molecular weight also affects the rheology, and data from synthetic polymers (Bin Ahmad & Ashby, 1988) show that the viscosity increases linearly with molecular weight below a critical value. Above this molecular weight a power law relationship holds:

$$\eta = C \, (M_r)^m \tag{6}$$

where M_r is the relative molecular mass.

The effect of concentration and composition in these systems has been included empirically, for example the relationship between shear viscosity and moisture content, M, has often been written in the form:

$$\eta = k \exp DM \tag{7}$$

and the effect of fat content in starch materials has also been included using an exponential term.

The effect of time has been neglected so far and clearly the flow properties will change with the progress of reactions. This factor is particularly difficult to include in the viscosity equation, although one attempt is due to Remsen & Clark (1978), who proposed that reactions which involve aggregation or network formation are analogous to increases in molecular weight. This is the basis for the incorporation of an extra term in equation (3):

$$\ln \eta = \ln k + \frac{\Delta E_\eta}{RT} + \int_{t_o}^{t} K' \exp \frac{\Delta E_k}{RT} dt \tag{8}$$

where E_k is an activation energy for reactions, E_η is an activation energy for flow, t is time and t_o is the start time for reaction. For example, the value of t_o could be established by the time when the temperature reaches a critical value for denaturation. The viscosity increases as part of the reduction of mobility required in materials modification and in the latter stages of product formation.

The shear viscosity is only one aspect of the flow properties. In finite length and non-parallel flows, the extensional viscosity is important and is linked to expansion in forming (Senouci & Smith, 1988b) such as the phenomenon of die swell in polymers, where the extrudate has a larger cross-section than the die and increases with increasing shear rate. Another elastic effect is melt fracture, which is a flow instability occurring above a critical shear stress.

A general indication of the properties of a viscoelastic material relative to a process is given by the Deborah number, De.

$$De = \lambda/t \tag{9}$$

where t is a characteristic time of the process and λ is a relaxation time of the material (Tadmor & Gogos, 1979). When the time of the process is very much shorter than the characteristic relaxation time of the material, the material will behave with solid-like properties; at low Deborah numbers, when the process time is much longer than the relaxation time, the material will behave with liquid-like characteristics. This general rule applies to 'processes' as diverse as eating and is consistent with the propensity to a brittle response at high deformation rates.

Materials – process interaction

The processing operations covered earlier are extremely complex and in general remain poorly understood. Ideally the process parameters and raw materials should be capable of specification to produce a product of given properties, including texture. Before this can be done the operations need to be defined in terms of the basic underlying transformations in engineering and physico-chemical terms including the occurrence of reactions and phase transitions. This in turn requires a knowledge of the structural, rheological, thermal and mechanical properties of food materials and their constituents in relation to strain, stress, temperature and time. In general, food components have been studied under ambient dilute conditions, although it is often the properties of concentrated, multi-component systems which are relevant to many processing operations.

Interaction of components

Food components interact with water in specific ways, as outlined above. In addition this interaction is affected by other components, for example the gelatinisation temperature of starch is raised by the addition of gluten, salt, fats and emulsifiers. The manipulation of the material ingredients to give the desired product properties has been illustrated in the context of baking (Donovan, 1977); sucrose is used to manipulate the protein denaturation and starch gelatinisation temperatures to occur nearly simultaneously at a temperature corresponding to maximum volume.

Baked products may be based simply on flour and water, the quantity and degree of development of gluten having a large effect on the extensibility of the material prior to structure formation. Fats and sugars are, however, included to extend the range of textures, appearance and taste; biscuits tend to have more sugar and fat and less water compared to bread. Fats act primarily to prevent the development of the gluten by coating the flour particles and may be added in some applications to compensate for the extra protein in a hard flour. Sugar disrupts the gluten network to reduce its stiffness and will also tend to harden the biscuit through solidification on cooling and as the water content falls. Abboud & Hoseney (1984) report that starch is not gelatinised on baking some biscuit doughs comprising 60% sugar. Brittle baked products may contain starch in swollen, gelatinised or disrupted form (Blanshard, 1986).

In a limited water system the components will compete for the water and this will change in favour of different constituents during processing. The resulting effects have been exploited in commercial processes to manipulate texture.

Product formation

The processing of materials to create a brittle product often involves: (1) mixing and preparation of a mobile material, (2) composite structure formation often involving introduction of a gaseous phase, (3) specific materials modification, and (4) removal of mobility including solidification. This is illustrated with reference to some of the processes referred to above.

1. Mixing may be carried out in a wide range of equipment and the nature of the strain, temperature and time history affects the process response and some of the product attributes. For many processes mixing results in a high viscosity dough or dispersion and affects protein structure, in particular gluten development. In baking of brittle products mixing takes place to hydrate the raw ingredients in a single operation or in a series of mixing or blending steps. Mixing leads to incorporation of air and foaming agents, which may begin to

release gases during mixing. It is intimately connected with rheological properties and hence the torque on a mixer is often monitored to indicate the progress of mixing.

The rheology of food materials is applicable to the preparation of the mobile material and provides an interface with the design of processing plant. The relationship between the rheology of the material and the texture of the final product has been cited in a number of cases. Frazier & Crawshaw (1984) found that the texture of extrusion-cooked defatted soya depended on the shear viscosity of the material in the die. In forming processes, the die pressure is one indicator of the rheology of the material and, in the case of extrusion-cooked foams, the prediction of densities from extrusion moisture content and die pressure has been demonstrated recently (Kirby *et al.*, 1988). In the early stages of baking there are significant changes in the rheological properties and the increase of the viscosity ultimately stops the expansion of included gas.

2. Baking powder breaks down under the influence of heat and moisture:

$$2NaHCO_3 \rightarrow Na_2CO_3 + H_2O + CO_2 \tag{10}$$

The solubility of carbon dioxide in water will decrease with increasing temperature and decreasing pressure. In some confections a sugar foam is produced by the inclusion of sodium bicarbonate, which decomposes to carbon dioxide at high temperature.

Foams are also produced by the incorporation of air through mixing or by directly injected gas where it is important to entrain as much air as possible to achieve highly expanded products. Gas bubbles may be incorporated for subsequent expansion, as in the case of pastries, the volume of gas increasing in proportion to temperature for a given pressure. Water expands 1600 times on vapourisation, expanding pastries, cereals and other starch products. Baked layered composites are produced by the construction of fat layers, which melt and form steam, which lifts the rest of the dough (Bloksma, 1986).

In other cases such as bread making and some baking applications the gas is produced by fermentation. Enzymes in yeast break down starch into sugar and sugars into carbon dioxide and ethanol:

Amylase   Maltase   Zymase

$$\text{Starch} \longrightarrow \text{Maltose} \longrightarrow \text{Glucose} \longrightarrow 2C_2H_5OH + 2CO_2 \qquad (11)$$

The yeast is inactivated at 55 °C.

3. This step is often highly process specific. Starch gelatinisation and protein coagulation increase viscosity and limit the expansion of gases as part of the structure formation. For example, in baking, the starch–protein framework must be viscous enough to trap gas bubbles formed on mixing and heating; sugar solutions also contribute to trapping gases particularly when the proteins do not fulfil this role. The matrix must also have the correct mechanical properties to retain this structure on solidification (step 4).

Processing may result in materials modification specifically as a result of the conditions and the state of the material. Degradation of food components takes place in concentrated, high viscosity systems with depolymerisation of constituent food polymers occurring, as in the case of dextrinisation in the extrusion cooking of starches (Mercier *et al.*, 1979).

Thermal and chemical degradation may also occur in most food components and structural modification is also likely in high shear conditions. Starch granules become distorted and eventually lose their identity in shear of highly viscous material (Richmond & Smith, 1987).

4. The final stage in product formation is usually cooling and drying, both of which serve to reduce mobility. Brittle properties are associated with low temperatures and low moisture contents; for example baked products may be described as composites comprising a polysaccharide–protein matrix filled by starch granules in which water and sugar act as plasticisers (Blanshard, 1986). Brittle properties of polymeric materials rely on the glassy state of the material, which may change with the diffusion of water and component interaction. Structuring, including foaming, may take place concurrently with mobility reduction.

The uniqueness of a particular processing route is brought into question by the example of flakes. Two processing routes may be used for certain breakfast cereals (Figs. 1 and 2); at the raw material stage extrusion cooking is able to handle a range of particle sizes including grits and flour. While one process subjects the material to a high temperature and high pressure, the other additionally shears the material. Holm, Björck &

Table 3. *Starch modification in processing of wheat (from Holm* et al., *1988)*

Technique	Starch granule	Water solubility index
Whole grain wheat		
Popping	Swollen – destroyed	3.5–40.3
Steam flaking	Partially gelatinised – swollen	0.7–1.7
Whole grain wheat flour		
Extrusion-cooking	Destroyed	40.9–41.4
Drum-drying	Destroyed	5.3–5.4

Eliasson (1988) compared a number of processing techniques for wheat and examined the starch characteristics together with the solubility of the product (Table 3). Even if the products of the two processes appear to be similar in terms of mechanical properties, density and porosity, it is the interaction of the product with cold or hot milk that affects consumer acceptance. The decay of brittle properties is therefore important and the time-dependent solubility needs to be known. The water solubility index of extrusion-cooked maize has been related to the specific mechanical energy input by the process in recent studies (Kirby *et al.*, 1988).

Drying

Drying by hot air consists of simultaneous transfer of (1) mass from the product to the air and within the drying substance itself and (2) heat from the heat source to the air, and from the air to the product and thence to the product interior.

Cellular materials
Fruit and vegetables may be considered as liquid-filled foams. Water filling the cells will tend to stabilise the cell walls against deformation, with turgor pressure being balanced by tensile stresses in the walls, which will also tend to raise the strength (Jeronimidis, 1988). In the processing of plant and animal tissue, some of the initial cellular structure is often preserved, as in blanching, freezing and drying of fruits and vegetables. The material may be considered to comprise a matrix which is usually water insoluble, intact or destroyed cell membranes and a complex aqueous solution of food constituents. The cell membranes influence the water changes depending on their degree of damage.

In a thermodynamic approach to water accounting, Crapiste & Rotstein (1982) considered the contributions to the chemical potential from the different elements of cellular materials. The chemical potential of water in each constituent is calculated on the basis that the number of moles of solute, n_s, of insoluble materials, n_m, and of water, n_w, are the relevant variables:

$$d\mu = VdP - SdT + \frac{d\mu}{dn_s} dn_s + \frac{d\mu}{dn_m} dn_m + \frac{d\mu}{dn_w} dn_w \qquad (12)$$

Integrating equation (12) with reference to pure water at atmospheric pressure gives:

$$\mu_w - \mu_{wo} = V(P - P_o) + RT \ln a_w + V_w \psi_m \qquad (13)$$

where μ_{wo} is a reference chemical potential at atmospheric pressure P_o and ψ_m is the integrated effect of the sorptional force field terms.

The first term in equation (13) represents the effect of intact membrane and the development of an internal pressure due to osmotic forces using a relationship between moisture content and turgor pressure or elastic behaviour of the matrix of insoluble material. The second term is the classical solution contribution and the third is the effect of the capillarity, adsorption and hydration forces. The internal and external atmospheres are assumed to be equal on the basis that intercellular air spaces are connected. Considering moist air to behave as an ideal gas:

$$\mu_w - \mu_{wo} = RT \ln \phi \qquad (14)$$

where ϕ is the water vapour activity.

At equilibrium and assuming no chemical reactions, equations (13) and (14) are equal (equation (1)). For a given value of ϕ the volume of water can be calculated in each constituent and these summed to give the water content. Crapiste & Rotstein (1982) showed the relationship for potato and its constituents. Starch is the main contributor to the total moisture content, although when ϕ approaches unity the vacuolar moisture content is dominant. This approach shows the contributions of the elements of a complex system and the shortcomings of the various correlations between moisture content and water activity, particularly at high moisture contents.

Freeze-drying

Compared to other drying techniques freeze-drying preserves the approximate shape of the material before freezing and yields a low bulk density structure. When the freeze-dried structure is heated, shrinking occurs, which is termed structural collapse, involving an irreversible loss of porosity. To & Flink (1978*a-c*) recorded collapse temperatures of $-10°C$ for starches and proteins and $-40°C$ for glucose and fructose. Under conditions of enhanced mobility the matrix relaxes so that the foam can no longer support its own weight. The freeze-dried product undergoes a brittle–ductile transition such that it may be deformed to higher strains without failure.

Formed products

The variation of the mechanical properties of pasta with moisture content have been measured by Andrieu & Stamatopoulos (1986) (Fig. 7). The shape of this plot is similar to that of Fig. 3 for a synthetic polymer as a function of temperature, which is to be expected if pasta is plasticised by water and its glass transition is lowered. The drying curves of pasta may be interpreted using a Fickian diffusional model and yield different diffusion coefficients depending on the moisture content range.

The density of ideal foams has been related to their mechanical properties (Gibson & Ashby, 1988):

$$\frac{\sigma}{\sigma_w} \propto \left(\frac{\varrho}{\varrho_w}\right)^n \tag{15}$$

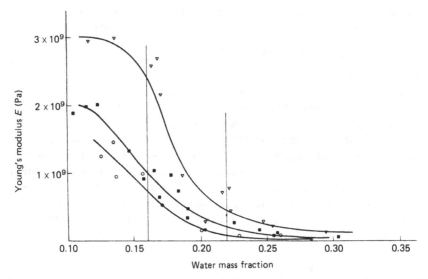

Fig. 7. The Young's modulus of durum wheat pasta as a function of mean solid moisture content (mass fraction). Pasta radius $= 1.5 \times 10^{-3}$ m. Temperatures: (∇) 20°C; ($\blacksquare$) 50°C; ($\bigcirc$) 70°C. (From Andrieu & Stamatopoulos, 1986, with permission.)

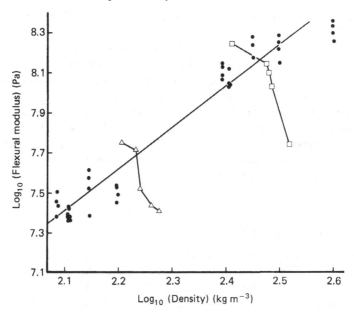

Fig. 8. The flexural modulus of wheat flour foams as a function of bulk density. (●) Variation of processing conditions; (□) moisture content conditioning.

where σ is the mechanical property of the foam and ϱ is its density and the subscript w refers to these properties for the foam wall material. In the case of extrusion-cooked foams, a good correlation has been found (Hutchinson, Siodlak & Smith, 1987) assuming constant wall properties. The importance of the wall properties is, however, clear in equation (15) and the wall mechanical properties will vary with strain rate, temperature and moisture content in much the same way as the bulk material. Complications may arise through orientation of the material during formation of the foam. The foam density may be manipulated by water sorption or desorption (Hutchinson, Mantle & Smith, 1989). Figure 8 shows the relationship between modulus and density for extrusion-cooked maize foams as processed, and also examples for which the density has been changed after extrusion.

This behaviour may be explained with reference to equation (15) which may be rewritten:

$$\frac{\sigma}{\sigma_w} \propto \left(\frac{V_w}{V}\right)^n \tag{16}$$

where V is the volume of the foam and V_w is the volume of the wall material.

Hydration of a foam will cause swelling, which will increase both V and V_w and, although there may be some increase in the ratio V_w/V, σ/σ_w will remain constant to a first approximation. Consequently σ will be proportional to σ_w. Water would be expected to plasticise the foam wall properties as it does to pasta (Fig. 7), which process would then be superimposed on the mechanical property–density power law to represent moisture uptake and loss (Fig. 8). A high density extruded foam which is then dried may have the same density as an extruded low density foam but the dried sample will have a greater strength and stiffness.

Product stability

The removal of water from materials may result in expansion, shrinkage or distortion. One of the problems in drying is the development of a moisture gradient in the product as a result of surface drying. In pasta drying this 'case-hardening' can trap water which turns to steam on further heating and escapes by breaking the surface layer. The products are then inferior on consumption, since they break apart and the interior material dissolves (Maurer, Tremblay & Chadwick, 1971). One solution is to dry in a humid environment, which is equivalent to drying more slowly in that it gives the material a greater mobility. Another approach is to adopt a direct drying technique, such as microwave or radio frequency, which maintains a more even moisture distribution.

Brittle products are subject to flaws which will reduce their strength such as the occurrence of cracks in biscuits called 'Checking', caused by freshly baked biscuits having a greater moisture content at the rim than at the centre. Moisture migration causes the development of stresses in the biscuit leading to spontaneous breakage (Wade, 1987), the solution being to reduce the temperature slowly to facilitate an even moisture distribution. The adoption of baking techniques which give an initially more even moisture distribution permits faster cooling of the biscuits.

The interaction of water and sugars is shown in the example of boiled sweets where the water content is minimised to stabilise the product against physical and chemical change. The low degree of mobility prevents aggregation and ordering into a crystal structure, although in a humid environment the product will absorb water which dilutes the surface sugar and crystallisation may then occur. Caking and agglomeration may also occur after drying.

Conclusions

The accommodation of temperature, time and moisture changes in polymeric food systems using the glass transition is attractive as a basis for the rational prediction of brittle textures, although the complete specification of processing conditions requires quantitative information that is largely unavailable. In very general terms processing may be seen as a means of raising the mobility of the system sufficiently to achieve mixing and flow followed by a reduction in mobility, using the material-specific properties in addition to cooling and solvent removal. Structure creation particularly in the production of foams must be integrated so that this occurs in the mobile system and is maintained on reducing the mobility.

References

Abboud, A.M. & Hoseney, R.C. (1984). Differential scanning calorimetry of sugar cookies and cookie doughs. *Cereal Chemistry*, **61**, 34–7.

Andrieu, J. & Stamatopoulos, A. (1986). Durum wheat pasta drying kinetics. *Food Science and Technology*, **19**, 211–14.

Atkins, C.J. (1968). *Equilibrium Thermodynamics*. McGraw Hill, London.

Bin Ahmad, Z. & Ashby, M.F. (1988). Failure mechanism maps for engineering polymers. *Journal of Materials Science*, **23**, 2037–50.

Blanshard, J.M.V. (1986). The significance of the structure and function of the starch granule in baked products. In *Chemistry and Physics of Baking*, ed. J.M.V. Blanshard, P.J. Frazier & T. Galliard, pp. 1–13. Royal Society of Chemistry, London.

Bloksma, A.H. (1986). Rheological aspects of structural changes during baking. In *Chemistry and Physics of Baking*, ed. J.M.V. Blanshard, P.J. Frazier & T. Galliard, pp. 170–8. Royal Society of Chemistry, London.

Brown, N. & Ward, I.M. (1983). The influence of morphology and molecular weight on ductile–brittle transitions in linear polyethylene. *Journal of Materials Science*, **18**, 1405–20.

Crapiste, G.H. & Rotstein, E. (1982). Prediction of sorptional equilibrium data for starch-containing foodstuffs. *Journal of Food Science*, **47**, 1501–7.

Donovan, J.W. (1977). A study of the baking process by differential scaning calorimetry. *Journal of the Science of Food and Agriculture*, **28**, 571–8.

Donovan, J.W. (1979). Phase transitions of the starch–water system. *Biopolymers*, **18**, 263–75.

Frazier, P.J. & Crawshaw, A. (1984). Relationship between die viscosity, ultrastructure and texture of extruded soya proteins. In

Thermal Processing and Quality of Foods, ed. P. Zeuthen, J.C. Cheftel, C. Eriksson, M. Jul, H. Leniger, P. Linko, G. Varela & G. Vos, pp. 89–95. Elsevier Applied Science, London.

Gal, S. (1983). The need for, and practical applications of, sorption data. In *Physical Properties of Food*, ed. R. Jowitt, F. Escher, B. Hallström, H.F.T. Meffert, W.E.L. Spiess & G. Vos, pp. 13–26. Applied Science, London.

Gibson, L.J. & Ashby, M.F. (1988). *Cellular solids – Structure and Properties*. Pergamon, Oxford.

Golden, J.H., Hammant, B.L. & Hazell, E.A. (1967). The effect of thermal pretreatment on the strength of polycarbonate. *Journal of Applied Polymer Science*, **11**, 1571–9.

Groeninckx, G., Berghmans, H. & Smets, G. (1976). Morphology and modulus-temperature behaviour of semicrystalline poly(ethylene terephthalate) (PET). *Journal of Polymer Science, Polymer Physics Edition*, **14**, 591–602.

Hiltner, A. & Baer, E. (1974). Mechanical properties of polymers at cryogenic temperatures: relationships between relaxation, yield and fracture processes. *Polymer*, **15**, 805–13.

Holm, J., Björck, I. & Eliasson, A.-C. (1988). Effects of thermal processing of wheat on starch. I. Physico-chemical and functional properties. *Journal of Cereal Science*, **8**, 249–60.

Hoseney, R.C., Zeleznak, K. & Lai, C.S. (1986). Wheat gluten: a glassy polymer. *Cereal Chemistry*, **63**, 285–6.

Hutchinson, R.J., Mantle, S.A. & Smith, A.C. (1989). The effect of moisture content on the mechanical properties of extruded food foams. *Journal of Materials Science*, **24**, 3249–53.

Hutchinson, R.J., Siodlak, G.D.E. & Smith, A.C. (1987). Influence of processing variables on the mechanical properties of extruded maize. *Journal of Materials Science*, **22**, 3956–62.

Jeronimidis, G. (1988). Structure and properties of liquid and solid foams. In *Food Structure – Its Creation and Evaluation*, ed. J.M.V. Blanchard and J.R. Mitchell, pp. 59–74. Butterworths, London.

Jones Parry, E. & Tabor, D. (1973). Effect of hydrostatic pressure on the mechanical properties of polymers: a brief review of published data. *Journal of Materials Science*, **8**, 1510–16.

Kirby, A.R., Ollett, A.-L., Parker, R. & Smith, A.C. (1988). An experimental study of screw configuration effects in the twin-screw extrusion cooking of maize grits. *Journal of Food Engineering*, **8**, 247–72.

Levine, H. & Slade, L. (1986). A polymer physico-chemical approach to the study of commercial starch hydrolysis products. *Carbohydrate Polymers*, **6**, 213–44.

Levine, H. & Slade, L. (1988). 'Collapse' phenomena – a unifying concept for interpreting the behaviour of low moisture foods. In *Food*

Structure – Its Creation and Evaluation, ed. J.M.V. Blanshard & J.R. Mitchell, pp. 149–80. Butterworths, London.

Lillford, P.J. (1986). Texturisation of proteins. In *Functional Properties of Food Macromolecules*, ed. J.R. Mitchell & D.A. Ledward, pp. 355–84. Elsevier Applied Science, London.

Lindley, M.G. (1988). Structured sugar systems. In *Food Structure – Its Creation and Evaluation*, ed. J.M.V. Blanshard & J.R. Mitchell, pp. 297–312. Butterworths, London.

Maurer, R.L., Tremblay, M.R. & Chadwick, E.A. (1971). Microwave processing of pasta. *Food Technology*, **25**, 1244–9.

Mercier, C., Charbonniere, R., Gallant, D. & Guilbot, A. (1979). Structural modification of various starches by extrusion cooking with a twin-screw French extruder. In *Polysaccharides in Food*, ed. J.M.V. Blanshard & J.R. Mitchell, pp. 153–70. Butterworths, London.

Muhr, A.H., Wetton, R.E. & Blanshard, J.M.V. (1982). Effect of hydrostatic pressure on starch gelatinisation as determined by DTA. *Carbohydrate Polymers*, **2**, 91–102.

Powell, P.C. (1983). *Engineering with Polymers*. Chapman and Hall, London.

Remsen, C.H. & Clark, J.P. (1978). A viscosity model for a cooking dough. *Journal of Food Process Engineering*, **2**, 39–64.

Richmond, P. & Smith, A.C. (1987). Rheology, structure and food processing. In *Food Structure and Behaviour*, ed. J.M.V. Blanshard & P.J. Lillford, pp. 259–83. Academic Press, London.

Senouci, A. & Smith, A.C. (1988a). An experimental study of food melt rheology. I. Shear viscosity using a slit die rheometer and a capillary rheometer. *Rheologica Acta*, **27**, 546–54.

Senouci, A. & Smith, A.C. (1988b). An experimental study of food melt rheology. II. End pressure effects. *Rheologica Acta*, **27**, 649–55.

Slade, L. & Levine, H. (1988). Structural stability of intermediate moisture foods – a new understanding? In *Food Structure – Its Creation and Evaluation*, ed. J.M.V. Blanshard & J.R. Mitchell, pp. 115–48. Butterworths, London.

Tadmor, Z. & Gogos, C.G. (1979). *Principles of Polymer Processing*. Wiley Interscience, New York.

To, E.C. & Flink, J.M. (1978a). 'Collapse', a structural transition in freeze dried carbohydrates. I. Evaluation of analytical methods. *Journal of Food Technology*, **13**, 551–65.

To, E.C. & Flink, J.M. (1978b). 'Collapse', a structural transition in freeze dried carbohydrates. II. Effect of solute composition. *Journal of Food Technology*, **13**, 567–81.

To, E.C. & Flink, J.M. (1978c). 'Collapse', a structural transition in freeze dried carbohydrates. III. Prerequisite of crystallization. *Journal of Food Technology*, **13**, 583–94.

210   A. C. SMITH

Van den Berg, C. (1986). Water activity. In *Concentration and Drying of Foods*, ed. D. MacCarthy, pp. 11–36. Applied Science, London.
Van Zuilichem, D.J., Tempel, W.J., Stolp, W. & Van't Riet, K. (1985). Production of high-boiled sugar confectionery by extrusion cooking of sucrose : liquid glucose mixtures. *Journal of Food Engineering*, **4**, 37–51.
Visser, J. (1988). Dry spinning of milk proteins. In *Food Structure – Its Creation and Evaluation*, ed. J.M.V. Blanshard & J.R. Mitchell, pp. 197–218. Butterworths, London.
Wade, P. (1987). Biscuit baking by near-infrared radiation. *Journal of Food Engineering*, **6**, 165–75.
Ward, I.M. (1983). *Mechanical Properties of Solid Polymers*. Wiley, Chichester.
Zeleznak, K.J. & Hoseney, R.C. (1987). The glass transition in starch. *Cereal Chemistry*, **64**, 211–14.

G. RODGER

The control and generation of texture in soft manufactured foods

To illustrate the origins of textures in a soft food, fish has been chosen as the example, since it is possible to show (1) how different process treatments applied to an unvarying raw material can produce different or similar textural properties, (2) how intrinsic and induced properties of the raw material can be responsible for different textures from a constant process, and (3) how novel products can be created based on the molecular properties of the raw material. As a result, this chapter is effectively divided into two main parts. The first describes three traditional fish product types which do not require an extreme alteration to the overall structural characteristics of the fish tissue; the second describes a product range where all native structure is broken down to gain access to the molecular properties of the materials comprising that structure.

Fish muscle structural organisation

Before dealing with those processes which affect texture but do not destroy the basic structural properties of fish muscle tissue, a description of that basic structure is necessary. The skeletal musculature of fish is segmented into a number of muscle blocks known as myotomes, which are separated from each other by encompassing sheaths of connective tissue called myocommata (Love, 1970). Each myotome comprises a large number of muscle fibres (in general shorter than those in mammalian muscle), which run obliquely to the main axis of the muscle system. Figure 1 shows a typical fish muscle fibre, while Fig. 2 shows the arrangement of the myotomes within a fillet (i.e. the musculature removed as a single cut from the side of a fish). Within each myotome, the fibres appear to be arranged into bundles surrounded by connective tissue sheaths thicker than the endomysia which surround each fibre. A natural inclination then exists to equate this to the hierarchy which has been identified in mammalian muscle (Bailey, 1983); in this equation the

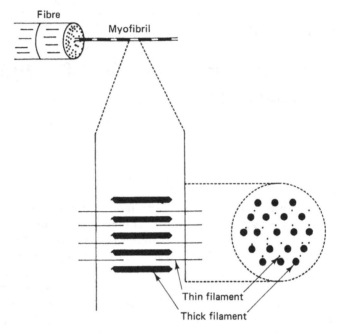

Fig. 1. The microstructure of a typical fish muscle fibre.

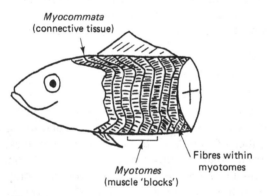

Fig. 2. A typical organisation of musculature in fish. (Adapted from Suzuki, 1981.)

myotome would be the equivalent of a mammalian muscle type, a view which is supported by Bremner & Hallett (1985).

As stated, the basic unit of muscle tissue is the muscle cell, or fibre, and as far as textural characteristics are concerned can be considered to consist of two complex components:

The contractile apparatus (the myofibrils).
The connective tissue which forms the cell envelope.

In fish, fibres are typically up to 3 cm long and 10–100 μm in diameter.

Myofibril

About 80% of the muscle tissue volume is occupied by the myofibrils, which are filamentous in nature (~ 1 μm in diameter) and run the length of the cell. The proteins which comprise the major part of this system are myosin and actin: their organisation in the myofibril is shown in Fig. 3, and illustrates the two sets of repeating filaments (thick and thin) which are aligned parallel to the long axis of the myofibril.

Under nearly all commercial applications of fish as food, the proteins actin and myosin exist as a complex protein actomyosin – i.e. those transitory linkages which can exist between the two molecules in conferring *in vivo* muscle function become fixed after death.

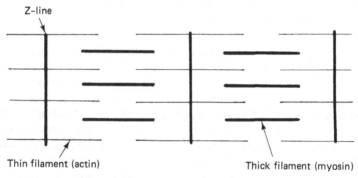

Fig. 3. Organisation of myosin and actin in fish skeletal muscle.

Connective tissue

By far the largest component in fish connective tissue is the protein collagen. Fish collagen differs considerably from mammalian collagen in that it contains far fewer cross-links and, on heating, a high proportion is converted to gelatin. In fish the collagen denatures and shrinks at an onset temperature of approximately 45 °C (Hastings *et al.*, 1985).

Traditional processes and the resulting product types

Salting and part drying (smoked salmon)

Smoked salmon is an example of a product which has evolved as a result of a food preservation process (salting/smoking/part-drying) enabling the raw material to be consumed in a relatively undisturbed physico-chemical state.

The process involves contacting the fish tissue with salt crystals, which allows diffusion of dissolved salt into the tissue until the required level is achieved. (It should be stressed that dry salting also removes some tissue water via osmotic influences and any excess extra-tissue water is allowed to drain away – this is in contrast to the immersion of muscle tissue in salt solutions (dilute), where the tendency is for the tissue to absorb water and swell.)

Once salted, the fish is 'cold-smoked', i.e. the salmon is allowed to absorb characteristic flavours from the smoke, and part-dried until the system's water activity (a_w) is low enough for a satisfactory shelf life to be achieved. The reader is referred to Mossel (1975) for a detailed description of the influence of a_w on the shelf-life of foods.

The sensory response to eating smoked salmon is similar to that evoked by a slightly chewy gel. There is little fibrosity in the product, and the overall succulence perception is moist but not wet. We can explain these effects by considering that fibrosity perception demands that the food product be mechanically anisotropic – i.e. that it exhibits a differential mechanical response to forces applied parallel or orthogonal to the muscle fibre direction. In smoked salmon it can be argued that this does not occur, since the connective tissue matrix has *not* been thermally modified, and therefore dominates the mechanical response of the tissue to applied force – i.e. the tissue is mechanically (though not visually) isotropic. The succulence behaviour (i.e. moistness remaining throughout chewing) can be explained by consideration of where water is present in the tissue. Because no heat processing has occurred, the majority of the water is still within the myofibrillar space – thus none is very easily expressible during the first few chews (it is physically trapped, essentially by capillary forces) and is only slowly released thereafter. This explanation does not, of course, exclude the contribution of tissue fat to the general succulence characteristics.

Thermal processing (e.g. fried, steamed, boiled fish)

Anyone who has ever cooked fish will have noticed that the application of heat to fish has a gross effect on the physico-chemical properties of the system. In most fish, except those with intrinsic colour, the appearance becomes milky white after being translucent in the raw state, and accompanying this change of colour is a visible shrinkage of, and loss of water from, the tissue. The mechanical properties of the muscle also alter, since the individual fibres comprising the structure become more rigid and yet the structure as a whole becomes somewhat weaker. These physico-chemical changes in the bulk material can be explained by events at the molecular level. The change in 'colour' results from thermal denaturation effecting tissue protein aggregation, which scatters light rather than allowing it to pass through the tissue. The influence of heat on the expulsion of water, and tissue shrinkage, occurs via the following sequence of events. As the tissue is heated, its temperature rises with little observable effect until the thermal denaturation temperature of myosin (about 37°C for most fish) is reached. The myofibrils begin to shrink as a result, causing (1) a spontaneous expulsion of myofibrillar water into the extracellular space, (2) an effective concentration of the myofibrillar protein, and (3) an increase in the fibre rigidity. As the temperature continues to rise the collagen in the connective tissue also begins to undergo thermal denaturation and shrinkage, which compounds the expulsion of water from the cell thus also increasing protein concentration. Finally, at about 65°C the actin component also denatures. This sequence of events, as followed by differential scanning calorimetry, is depicted in Fig. 4.

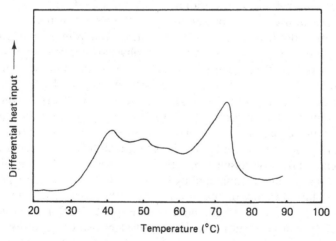

Fig. 4. A differential scanning calorimetry profile of the thermal denaturation of the proteins in cod muscle. (Adapted from Hastings *et al.*, 1985.)

The reason why individual fibres become more rigid is adequately explained by the effects of heat on the myofibrillar proteins, but the decrease in the strength of the structure as a whole depends on the connective tissue. This effect is very noticeable in fish tissue, since fish collagen possesses few heat-stable cross-links and as a result readily undergoes conversion to gelatin, which has low mechanical strength. The net effect is the ready breakdown of cooked fish muscle into flakes (the primary level of connective tissue organisation) and then into fibres (the secondary level of organisation), on the application of even very gentle force.

The previous assertion that fibrosity perception demands mechanical anisotropy, i.e. preferential planes of weakness in the tissue, is met as a result of cooking, since the fibres become stronger and the connective tissue weaker. Thus, when the forces exerted by the mouth during chewing are applied to cooked tissue, the structure preferentially fails at the connective tissue level and the muscle fibres can then be sensed. The other sensory characteristics which depend on the mechanical properties of the tissue, i.e. toughness, chewiness, etc., are dependent only on the fibre properties of fish muscle. As far as the author is aware, connective tissue contributes very little, if anything, to the sensory texture of thermally processed fish.

High ionic strength/low pH (e.g. marinaded herring)

Marinaded herring is a traditional food consumed in several North European countries where preservation is effected by curing the fish in a solution of sodium chloride and acetic acid. Inherent in the process, however, are the changes in both the physico-chemical and sensory properties of the tissue caused by the increase in ionic strength and decrease in pH. These changes have been investigated and described (Rodger et al., 1984), but the origins of the effects observed when fish muscle is marinaded are generally as follows. When we consider that the proteins which constitute muscle have functional groups which respond to changes in ionic strength and pH, it is not difficult to see why salt and acid have such an effect on tissue properties. The response of the myofibrillar proteins to acid can be explained by referring to their isoelectric point. If the pH is shifted in any direction away from the isoelectric point (IEP), then the proteins attain either a net negative (pH > IEP) or a net positive (pH < IEP) charge resulting in increased electrostatic repulsion of the proteins. The result of this is that the structure can swell and more water can be held. Since the pH of fish flesh post rigor (6.2–6.8) is greater than the IEP, the progressive addition of acid to pH 4 should see the structure

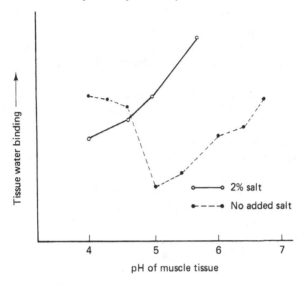

Fig. 5. The influence of pH, in the presence, and absence of salt, on the water-holding ability of herring muscle tissue. (Adapted from Rodger *et al.*, 1984.)

shrink as the pH approaches the IEP (~ 5.4) and then begin to swell again. Fig. 5 shows that this does in fact occur.

However, if salt is present, then the pH at which the water binding of the system is minimal is shifted to a lower pH value (~ 4), which is much nearer the equilibrated pH of the marinade. Thus, in the presence of salt and acid *together*, the physico-chemical response of the tissue is different from the response to either salt and acid alone.

An additional point worth mentioning is that if fish are immersed in acid alone, they become exceedingly fragile, since the connective tissue which holds the myotomes together is soluble in dilute acid at low ionic strength.

The sensory properties of the tissue brought about by the marinading process can be explained from the physico-chemical effects described above. The pH and ionic strength of the system change as acid and salt diffuse into the tissue. Since the IEP of the tissue at higher ionic strength decreases, the muscle fibres shrink and expel water to the extracellular space. The fibres become firmer, since water expulsion increases myofibrillar protein density. At the same time, the connective tissue becomes weaker (although the presence of the salt prevents the dissolution which occurs in acid alone). The net effect is that the system becomes mechanically anisotropic – on the application of chewing forces, it preferentially fails at the connective tissue level and sensory fibrosity develops as a result. In addition, because of the expulsion of water to the extracellular space, where it is less firmly 'bound' by capillary forces (the pores are bigger), the marinaded product is wetter during the first few chews than the untreated tissue would be.

Marinaded products are, however, subjected to influences other than the direct effects of salt and acid. During storage of the product, the perceived firmness changes. As the salt and acid diffuse into the tissue, the firmness increases for the reasons described above. When the system is 'equilibrated' with respect to pH and ionic strength, there is a period of relative 'stability', after which the tissue softens. This latter effect has been assigned to the acid activation of a group of proteolytic enzymes called cathepsins, which attack the myofibrillar proteins, thereby weakening the overall structure.

In textural (*not* flavour) terms, the sensory response of marinaded herring is very similar to the heat-processed product, and in this sense marinading can be considered as 'chemical-cooking'.

State of raw material

The previous sections have outlined the effects which three different processes have on the sensory properties of fish muscle tissue. However, the state of the raw material used in the process can also contribute significantly to the product properties, irrespective of subsequent processing. The factors which contribute to raw material state (or 'quality') can usefully be described as follows:

Intrinsic influences

These are the influences on the system over which the product manufacturer has little control (e.g. state of rigor, size, nutritional state, seasonality, catching ground), but it must be remembered that in most cases, fish are still a hunted food. However, seasonality or catching-ground influences do not explain the physico-chemical origins of the effect. In general, intrinsic effects in fish can adequately be described by consideration of pH and water content for white fish, plus the influence of fat content for herring, salmon etc., as follows:

In fish, 'seasonality' and 'catching ground' are related to the nutritional state of the fish via the spawning event (spawning occurs at different times in different grounds). During the period, fish metabolise protein as an energy source – consequently the tissue water content increases as the protein content decreases and depleted glycogen levels mean that the

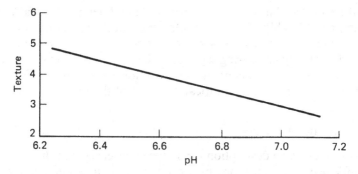

Fig. 6. The influence of post-mortem pH on the texture of cooked cod muscle (constant H_2O composition): > 3, tough/firm; < 3, soft. (Adapted from Love *et al.*, 1974.)

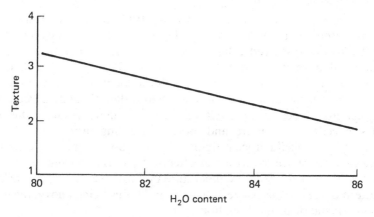

Fig. 7. The influence of water content on the texture of cooked cod muscle (constant pH). (Adapted from Love *et al.*, 1974.) For texture values, see Fig. 6.

post mortem pH is high. The net physico-chemical effect is that individual fibres become mechanically less strong (decreased protein concentration) and thus relative to a non-spawning fish, the cooked product is softer. Love *et al.* (1974) showed this pH dependence as well as showing that the tissue water content can also influence perceived sensory texture (Figs. 6 and 7).

The mechanistic effect of pH on texture is as described under High ionic strength, above. As the pH moves toward the IEP of the muscle (at constant salt level), the tissue shrinks and becomes mechanically stronger, via protein concentration effects.

In the so-called fatty species, e.g. salmon and herring, the fat content can vary considerably with season. One major influence of the amount of deposited fat is on the overall succulence perception – low fat herring are firm and relatively dry during successive chews, while high fat fish tend to be slightly softer, but maintain the sensation of succulence for longer during mastication, via the lubricating effect of the fat.

Induced effects

The best alternative description for 'induced' effects is post-harvest handling influences. In the fish-processing industry, the major effect on texture arises via the freezing and subsequent frozen storage of the fish. (Sensory effects which are a result of *bad* handling, e.g. allowing freezer-burn to develop during frozen storage or microbial spoilage to occur during wet storage are not considered here.)

When fish muscle tissue is cooled below 0 °C, ice does not usually begin to form until −2 to −4 °C is reached because of the freezing point depressant effect of dissolved cellular solutes. As the temperature decreases, more and more water is frozen. In most commercial freezing operations, the rate of freezing is generally not fast enough to effect intracellular freezing and, therefore, ice nucleates extracellularly and subsequent ice crystal growth occurs there (Calvelo, 1981). The net result of this is that muscle cells shrink more and more as freezing progresses, since the growth of extracellular ice promotes the transport of cellular water to the extracellular space via osmotic effects. If, during frozen storage, the muscle tissue experiences temperature cycling, then the size distribution pattern of ice crystals is altered significantly, this being known technically as ice crystal disproportionation.

In fish muscle, the effects of frozen storage can be extreme, and can depend on the physical state of the muscle at the time of freezing (i.e. whether the muscle is intact, e.g. whole fish/fillets or in the form of a coarse comminute, e.g. commercially available fish mince) and the temperature and time of frozen storage. In terms of the effect on the bulk properties of fish muscle, the most obvious is the altered water distribution; on thawing frozen-stored fish, there can be copious loss of cellular water in the form of 'drip'. In addition, the tissue itself becomes firmer and more elastic. We consider that the most likely explanation for this is that the freezing process (by inducing water redistribution) effectively concentrates the myofibrillar protein within the muscle fibre. If the tissue is frozen-stored for only a short time, most of this redistributed water is re-imbibed by the fibres. If, however, there is a prolonged storage of the tissue, an as yet not fully understood physico-chemical reaction manifests itself as the oft-quoted freeze-denaturation mechanism. For a full description of the possible causes of this phenomenon, the reader is referred to Shenouda (1980). The best supported theory is that in the fish family known as Gadidae an enzyme-mediated reaction converts trimethylamine into dimethylamine plus formaldehyde. The latter compound then effects a cross-linking between adjacent myofibrillar proteins. This fixes the proteins in their freeze-concentrated state so that on thawing of the tissue the fibres are unable to swell and regain their lost water. Fortunately, this explanation can also accommodate the changes which occur in the mechanical properties of the tissue; the increased concentration and chemical fixing of the myofibrillar proteins would lead to the fibres becoming more rigid. There is strong evidence, from proton transverse relaxation time (T_2) nuclear magnetic resonance studies, to support the assertion that water redistribution is one of the prime events occurring during frozen storage of fish. Such a study (Lillford, Jones & Rodger, 1980) has shown that prolonged frozen-storage of coarsely comminuted

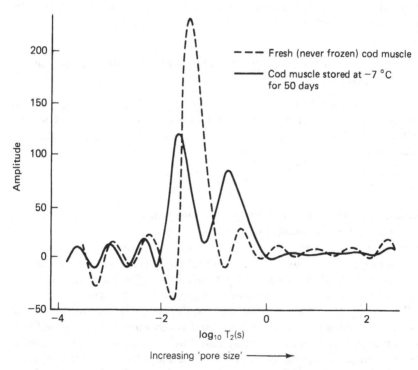

Fig. 8. T_2 nuclear magnetic resonance relaxation analysis of fresh and frozen cod muscle. (Adapted from Lillford *et al.*, 1980.)

fish induces irreversible structural alterations. These structural changes are considered to arise from aggregation of the myofibrillar components during the storage period. As a consequence of this, sufficiently large spaces in the system are created so that exchange of 'bound' and 'free' water cannot be averaged over the complete sample. Thus, 'freeze-denaturation' of fish muscle can then be explained simply in terms of increasing aggregation, resulting in an increase in the number or size of interstitial spaces (Fig. 8). An additional consequence of this aggregation or freeze-denaturation of fish muscle is that the solubilities of the constituent myofibrillar proteins are dramatically changed. From being readily soluble in 0.6 M NaCl solution they can become virtually insoluble after frozen storage, but this is very much dependent on storage time and temperature.

The influence of these physico-chemical changes on the properties of the raw muscle tissue is *not* removed by heat treatment. In other words, the increased rigidity in fibres effected by frozen storage/pH decrease carries through to the cooked product made from these raw materials, i.e. they will be firmer and 'tougher' than the non-frozen-stored/high pH raw material cooked under identical conditions.

Products based on gelling characteristics of fish muscle proteins

Fine comminution of fish muscle to attain the gel-forming properties of the muscle actomysin has been used in Japan for centuries. The raw material which results from the process and is commercially available is called surimi.

Surimi

Surimi is, in the simplest terms, a homogeneous, concentrated form of purified fish muscle myofibrillar proteins, supplied for further processing in the form of frozen blocks. The preparation of surimi from whole fish is shown in Fig. 9. A general compositional analysis of this material before cryoprotectant addition (on a dry basis) is 90–95% protein. On a wet basis, the analysis is approximately 20% protein, 80% water.

The range of products made from surimi is extremely large, and in 1985, world production of surimi-based foods amounted to about one million tons, worth approximately £2400 million (Fretheim *et al.*, 1988). For the purposes of this chapter, only the manufacture of the 'fibrous' muscle tissue analogues (which account now for ~10% of world sales) will be considered. (This product type is currently marketed in the UK as

Table 1. *Formulation ingredients of surimi-based crab-sticks*

Surimi Water Salt	Essential for texture development
Starch Egg albumen Natural crab tissue	Optional for texture development
Crab flavour Flavour enhancer	Essential for flavour perception

Data adapted from Freitheim *et al.*, 1988.

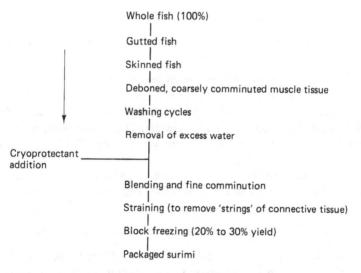

Whole fish (100%)
|
Gutted fish
|
Skinned fish
|
Deboned, coarsely comminuted muscle tissue
|
Washing cycles
|
Removal of excess water

Cryoprotectant
addition
|
Blending and fine comminution
|
Straining (to remove 'strings' of connective tissue)
|
Block freezing (20% to 30% yield)
|
Packaged surimi

Fig. 9. A typical preparation of surimi from whole fish.

'crab sticks'.) A typical ingredients list for such a product is given in Table 1, and the method of manufacture is shown in Fig. 10.

Surimi-based crab-sticks

To generate texture in this product type, two main events must occur. One is the creation of a protein gel and the other is the superimposition of the structure. In the manufacture of crab-sticks, this is achieved as follows. The addition of salt to the formulation ensures a degree of

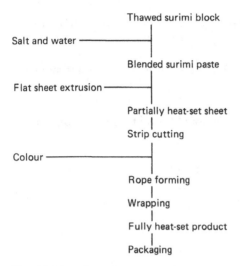

Fig. 10. Manufacturing process for textured surimi product (crab-sticks).

solubilisation of the myofibrillar proteins, which then partially gel during the first heating stage in the process. This allows the creation of the directional structure by the strip cutter, which introduces 'fracture planes' along the long axis of the product. The second heating stage then completes the thermal denaturation of the myofibrillar proteins and creates a stable structure.

The properties of these gelled systems can be influenced by the other ingredients listed in Table 1, but how they function will be discussed later.

Surimi – control of intrinsic raw material properties

As mentioned above, the quality of surimi depends on the molecular properties of its constituent myofibrillar proteins, principally the ability to form a firm gel on heating. 'Control' of this property is exerted in four main ways.

The first is the choice of the fish species to be processed, since several workers have shown that different species give widely different gel-forming capabilities. The various species which have been studied have been reviewed by Suzuki (1981, pp. 89–105). The technology supporting surimi production has in general developed using 'white' rather than 'fatty' fish (white = lean) for the following reasons:

1. A better quality raw material is obtained, since intramuscular fat interferes with the gel-forming process,

and can lead to flavour problems (rancidity development) during storage.

2. Process yields are higher.

The most common species used at present is the Alaska pollack (*Thelagra chalcogramma*), but, as more Western countries have become interested in this technology, those species more commonly landed, e.g. cod (*Gadus morhua*), capelin (*Mallotus villosus*), saithe (*Gadus virens*) and blue whiting (*Micromesistius poutassou*) have all been studied.

When the fish are harvested is also a point of primary control, since the quality of surimi varies with the season. The ideal time is when the fish are well fed, since the water content of muscle is lower and the higher glycogen levels in the tissue result in a pH which is lower (as a result of anaerobic glycolysis forming lactic acid) than that derived from post-spawning fish, which have depleted glycogen levels. As mentioned previously, a muscle protein system at high pH can absorb and 'bind' more water than at low pH, which decreases the efficiency of the de-watering process during surimi manufacture. It should be stated of course that the effect of pH can be used as a positive control point, since the properties of the raw material can be influenced by judicious acidification during processing. The influence of final water content on the gelling properties of surimi is via the effect on protein concentration, but pH manifests its effect both via its influence on water content (hence protein concentration) and the molecular properties of the proteins which comprise the gel (Yasui, Ishioroshi & Samejima 1980).

The second method of controlling the raw material contribution is to remove the water-soluble cytoplasmic proteins in the tissue by extensive washing with water. If these proteins are not removed, their presence reduces the ultimate achievable gel strength in two ways:

1. They 'dilute' the stronger-gel forming myofibrillar proteins.
2. They contain the enzyme system which can convert trimethylamine oxide to dimethylamine and formaldehyde. As described above, this can lead to the loss of protein solubility which, in turn, dramatically reduces the system's ability to form a gel (Kostuch & Sikorski, 1977).

As well as improving the textural quality of the final product, washing of the raw material also removes haem compounds which may catalyse lipid oxidation (sensory off-flavour) and darken (a disadvantage) the final colour of the product.

A third point which can influence raw material quality is post-harvest handling of the landed fish. If proteolytic enzymes leak out of the gut into surrounding muscle tissue (Su, Lin & Lanier, 1971*a,b*) then surimi prepared from that muscle tissue has lower gel strength than would normally be expected. The degree of leakage is time and temperature dependent and, in general, whole fish held for longer than 4 days at 0°C prior to processing do not yield good-quality surimi (R. K. Lippincot & C. M. Lee, unpublished data). This 4 day limit will obviously become less if the holding temperature is >0°C.

The fourth, and probably most important, factor which affects surimi quality is the freeze-denaturation of muscle proteins. Although the use of the word 'denaturation' is probably not strictly correct (there is no very convincing evidence that the myofibrillar proteins undergo significant conformational change), the result of the phenomenon is marked. The solubility of the proteins in a salt environment is much reduced and, since solubility is required for optimal gel formation, products manufactured from such a 'denatured' material do not exhibit the required gel properties of firmness and elasticity which the Japanese call 'ashi'.

The mechanism behind solubility loss is probably one, or both, of the following:

1. The cross-linking of proteins by formaldehyde generated from the conversion of trimethylamine oxide to dimethylamine and formaldehyde by trimethylamine oxidase as mentioned above.

2. The concentration-induced aggregation (with resultant solubility loss) of myofibrillar proteins resulting from the removal of water from the system as ice during frozen-storage.

The means of controlling this effect involves the use of cryoprotectants to prevent or minimise this solubility loss, but how cryoprotectants exert their influence is not yet fully understood.

As well as using cryoprotectants to maintain surimi quality, close attention must also be paid to storage temperatures. A maximum of −20°C is recommended, with as little temperature cycling as possible (Iwata et al., 1968, 1971); storage at −10°C yields a material which cannot be used for the manufacture of any high quality product (Iwata et al., 1971).

When surimi is processed into the final product, time–temperature relationships in the process become important. Fish muscle proteins in surimi exhibit the phenomenon (known as 'suwari') of undergoing a low temperature gelation at 40°C or below. If incubated at 40°C prior to full cooking at 90°C, then the textural strength of the gel is significantly greater than if no pre-incubation had occurred (Lanier, 1986). No attempt is made here to explain the cause and effect relationship, but the reader is referred to Montejano, Hamann & Lanier (1984), Niwa et al.

(1982; Niwa, Matsubara & Hamada, 1982; Niwa, Nakayama & Hamada, 1983), and Wu *et al.* (1985) for futher information.

Another time–temperature effect relates to the possible presence of proteolytic enzymes in the surimi. These, where they exist, exhibit temperature optima at 60–70 °C (Lin & Lanier, 1980; Lin, Su & Lanier, 1980) and, if for some reason surimi experiences a holding time in this temperature, then serious textural defects can be expected to arise.

Earlier, it was stated that ingredients other than the surimi component in the product formulation could influence final sensory quality via their effects on the gel properties. The addition of egg albumen creates a firmer gel primarily because the net protein concentration is increased, since egg albumen forms a gel of strength similar to that of the myofibrillar proteins, no 'dilution' of the gel strength results as is the case when muscle cytoplasmic proteins are included.

The starch (ungelatinised), which is added as granules, absorbs water from its environment during heating, which effectively concentrates the surrounding protein and causes the formation of a firmer gel. In addition, the swollen granules may also act as a 'reinforcement filler' in the newly created gel composite.

No evidence is currently available as to what influence varying the dimensions of the strips has on the resultant sensory quality, but, according to Lillford (1986), the dimensions of fibres created in a process should match those of the tissue food being simulated. On this basis, it seems realistic to suppose that control of the strip cutter width could yield a variety of final textures.

Summary

This chapter has attempted a description of some of the raw material properties ←→ process interactions which can give rise to a range of sensory textures in a soft-eating food product, in this case fish.

The examples chosen have tried to illustrate the concept that sensory texture is an integrated sensation of the forces required to reduce the food to a swallowable state, plus the sensation of fluid in the mouth en route to this state. Thus, texture can be considered to derive in part from both the food structure and the properties of the materials comprising that structure.

References

Bailey, A.J. (1983). The chemistry of intramuscular collagen. In *Recent Advances in the Chemistry of Meat*, ed. A.J. Bailey, pp. 22–40, *Royal Society of Chemistry Special Publication*, No. 47.

Bremner, H.A. & Hallett, I.C. (1985). Muscle fibre–connective tissue junctions in the fish blue grenadier – a scanning electron microscopy study. *Journal of Food Science*, **50**, 975–80.

Calvelo, A. (1981). Recent studies on meat freezing. In *Developments in Meat Science*, vol. 2, ed. R. Lawrie, pp. 125–58. Applied Science Publishers, London.

Fretheim, K., Egelandsdal, B., Langmyhr, E., Eide, O. & Ofstad, R. (1988). Surimi based foods – the general story and the Norwegian approach. In *Food Structure – Its Creation and Evaluation*, ed. J.M.V. Blanchard & J.R. Mitchell, pp. 265–77. Butterworths, London.

Hastings, R.J., Rodger, G.W., Park, R., Matthews, A.D. & Anderson, E.M. (1985). Differential scanning calorimetry of fish muscle: the effect of processing and species variation. *Journal of Food Science*, **50**, 503–6 and 510.

Iwata, K., Kanna, K., Umemoto, S. & Okada, M. (1971). Study of the quality of frozen stored Alaska pollack surimi. The influence of freshness of the material and changes in the storage temperature. *Bulletin of the Japanese Society of Scientific Fisheries*, **37**, 626–33.

Iwata, K., Okada, M., Fujii, Y. & Mimoto, K. (1968). Influences of storage temperatures on quality of frozen Alaska pollack surimi. *Reito (Refrigeration)*, **43**, 1145–8.

Kostuch, S. & Sikorski, Z.E. (1977). Interaction of formaldehyde with cod proteins during frozen storage. *International Institute of Refrigeration Communications C1 and C2*, Karlsruhe, pp. 199–208.

Lanier, T.C. (1986). Functional properties of surimi. *Food Technology*, **40**, 107–000.

Lillford, P.J. (1986). Texturisation of proteins. In *Functional Properties of Food Macromolecules*, ed. J.R. Mitchell & D. Ledward, pp. 355–84. Elsevier Applied Science, London.

Lillford, P.J., Jones, D. & Rodger, G. (1980). Water in fish tissue – a proton relaxation study of post rigor minced cod. In *Advances in Fish Science and Technology, Conference Edition*, ed. J.J. Connell, pp. 495–7. Fishing News Books Ltd., Farnham.

Lin, T.S. & Lanier, T. (1980). Properties of an alkaline protease from the skeletal muscle of Atlantic croaker. *Journal of Food Biochemistry*, **4**, 17–28.

Lin, T.S., Su, H.K. & Lanier, T.C. (1980). Characterisation of fish muscle proteins using radio-labelled protein substrates. *Journal of Food Science*, **45**, 1036–9.

Love, R.M. (1970). *The Chemistry and Biology of Fishes*, vol. 1. Academic Press, London & New York.

Love, R.M., Robertson, I., Smith, G.L. & Whittle, K.J. (1974). The texture of cod muscle. *Journal of Texture Studies*, 5, 201–12.

Montejano, J.G., Hamann, D.D. & Lanier, T.C. (1984). Thermally induced gelation of selected comminuted muscle systems; rheological changes during processing, final strengths, and micro-structure. *Journal of Food Science*, 40, 1496–505.

Mossel, D.A.A. (1975). Water and micro-organisms in foods – a synthesis. In *Water Relations of Foods*, ed. R.B. Duckworth, pp. 347–61. Academic Press, London.

Niwa, E., Matsubara, Y. & Hamada, I. (1982). Hydrogen and other polar bondings in fish flesh gel and setting gel. *Bulletin of the Japanese Society of Scientific Fisheries*, 48, 667–70.

Niwa, E., Matsubara, Y., Nakayama, T. & Hamada, I. (1982). Participation of S–S bonding in the appearance of setting. *Bulletin of the Japanese Society of Scientific Fisheries*, 48, 727–8.

Niwa, E., Nakayama, T. & Hamada, I. (1983). The third evidence for the participation of hydrophobic interactions in fish flesh gel formation. *Bulletin of the Japanese Society of Scientific Fisheries*, 49, 1763–5.

Rodger, G.W., Hastings, R., Cryne, C. & Bailey, J. (1984). Diffusion properties of salt and acetic acid into herring and their subsequent effect on the muscle tissue. *Journal of Food Science*, 49, 714–20.

Shenouda, S.Y.K. (1980). Theories of protein denaturation during frozen storage of fish flesh. *Advances in Food Research*, 26, 275–311.

Su, H., Lin, T.S. & Lanier, T.C. (1981a). Investigation into potential sources of heat-stable alkaline protease in mechanically separated Atlantic croaker. *Journal of Food Science*, 46, 1654–6.

Su, H., Lin, T.S. & Lanier, T.C. (1981b). Contribution of retained organ tissues to the alkaline protease of mechanically separated Atlantic croaker. *Journal of Food Science*, 46, 1650–3.

Suzuki, T. (1981). *Fish and Krill Protein: Processing Technology*. Applied Science Publishers, London.

Wu, M.C., Akahane, T., Lanier, T.C. & Hamann, D.D. (1985). Thermal transition of actomyosin and surimi prepared from Atlantic croaker as studied by differential scanning calorimetry. *Journal of Food Science*, 50, 10–13.

Yasui, T., Ishioroshi, M. & Samejima, K. (1980). Heat induced gelation of myosin in the presence of actin. *Journal of Food Biochemistry*, 4, 61–78.

P. J. LILLFORD

Texture and acceptability of human foods

Preference (acceptability) and texture perception are judgments made by each of us every time we eat, without much conscious thought. Two groups of experimentalists are interested in how the two are mechanistically linked. They are behavioural psychologists and food product designers. From behavioural studies we can draw a few simple conclusions.

First, eating is not an activity to which a great deal of analytical thought or concentration is normally applied. People behave as if their actions are 'scripted', i.e. they are acting out a process during which a sequence of events is to be expected (Abelson, 1981). Only if the unexpected occurs is any judgment logged.

Second, because of the scripted procedures, acceptability of food is dependent on the description or expectation of the properties of the food being eaten. For example, a simple sugar glass can be fabricated into a boiled sweet (hard) or an aerated structure (crunchy). The one is not normally an acceptable form of the other. Fortunately for the confectionery industry, both are acceptable food concepts if properly described. Conversely even if the texture and flavour of two products are closely matched, acceptability will not be obtained unless the concept is acceptable. For example, it has been found that well-matched but unlabelled samples of chicken meat and vegetable protein analogue are equally acceptable, but when the consumer is told which is which, the analogue is rejected because the idea of copying a 'natural' product is unacceptable.

Finally, there are some forms of highly nutritious materials which are almost inedible because of their incompatibility with the mechanics of the mouth. Examples are: the meat from the carcase of an old beef animal, which cannot be fractured in the mouth without treatment by cooking or comminution; most seeds which are difficult to chew and would be almost indigestible without milling and restructuring to the products of the bakery industry. The success of *Homo sapiens* must in part be due to our ability to adapt, by processing, materials which would not otherwise be consumable or nutritious.

Of course, the modern food-processing industry has added other desirable or acceptable qualities to raw materials, such as their convenience, 'image' and value for money which, while equally important, are not relevant to this volume.

When consumer research on food products is carried out, the simple question 'Do you like it?' is frequently asked. This can be quantified on an hedonic scale, but will include assessment of all the perceived attributes compared against expectation. Even untrained consumers are capable of separating their sensory judgments. Hence 'Do you like the texture?' is a similarly quantifiable judgment which frequently weighs heavily in the acceptability of a product.

So texture is important, not only because commercial companies survive by its control, but also because the human masticatory apparatus has evolved to cope with particular types of mechanical properties, and certain traditional foods are recognised by their mechanical properties. So texture is important, but what is it?

Food texture – some definitions and deductions

In common parlance, texture refers to the organisation of the constituent parts of a structure, normally detectable by eye. This is true also for foods. Experiments with meat and spun fibre analogues show that while both are described as 'fibrous', the analogue is described as less 'meat like' because of the uniformity of its fibre diameters and its non-plastic response to cutting into chunks. Perception and preference of texture appears to be influenced by the structural organisation at the visible level.

Perhaps more surprisingly panellists are prepared to make judgments on the 'moistness' of cakes simply by inspection of the cut surfaces. In control experiments where visual judgments were scored and later compared with in-mouth assessment of the same parameter, the correlation between visual only and in-mouth perception of moistness for the same structures was remarkably high. We may hypothesise that the visual judgments rely on previous experience, in which the pore sizes and sheen of the cut surfaces are normally correlated with the absorption of saliva and the barrier properties of fat in the formulation.

It is evident from the above that the visual appearance of texture may set the 'script' by which subsequent in-mouth textural judgments are made. These need not only be observations of surface structure and texture; visual colour and aroma emanating from or associated with foods can also influence the 'script' of expectation. Whether we like it or not, humans have reflexes conditioned in ways not dissimilar to those induced by Pavlov in his salivating dogs. Now we are close to the psychological aspects of advertising.

Returning to texture perception in the mouth, Ball, Clauss & Stier (1957) recognised the need to separate 'sight' and 'feel' definitions of meat texture. The former is related to fibrousness, as mentioned above. The 'feel' definition was simply stated as 'The smoothness or fineness of muscle tissue in the mouth'. Earlier workers had limited texture to the visible attributes only and preferred to use 'consistency' (Smith, 1947) or 'body' (Davis, 1937) as the mechanical in-mouth property.

In a classic paper in 1963, Szczesniak attempted to produce a unifying description of texture by compiling dictionary and rheological definition of popularly used terms. Her conclusion was that textural characteristics could be grouped into three main classes.

1. Mechanical.
2. Geometric.
3. Other – mostly relating to moisture and fat content, (lubrication).

The detailed relationships which she proposed between parameters and popular nomenclature are shown below (Table 1).

Szczesniak, Brandt & Friedman (1963) attempted to relate parameters to physical methods of measurement, and the texturometer was established. The classifications in Table 1 also served as the basis for a descriptor set of sensory parameters used for training panellists in sensory assessment.

Descriptors, attributes and mapping of textural space

Since the pioneering work of Szczesniak, a number of authors have developed sensory profiles of food texture and flavour using attributes or descriptors of the food agreed by panellists related to particular in-mouth sensations (Harries, Rhodes & Chrystall, 1972). It was soon recognised that for a particular type of food, such as meat, group discussion with the panellists produced a more detailed description of texture than the Szczesniak attributes could cope with. The problem was then to decide which should be used to obtain a complete description or at least to discriminate between qualities of closely related food types. Statistical relationships and analysis of data sets has been extremely useful but must be carefully applied. An early example of the procedure was demonstrated by using the sensory profile to predict the preference rating of rice (Schutz & Damrell, 1974). This model used stepwise regression to produce a regression equation using five sensory parameters. However, since

Table 1. *Relations between textural parameters and popular nomenclature (from Szczesniak* et al., *1963)*

Mechanical characteristics

Primary parameters	Secondary parameters	Popular terms
Hardness		Soft→Firm→Hard
Cohesiveness	Brittleness	Crumble→Crunchy→Brittle
	Chewiness	Tender→Chewy→Tough
	Gumminess	Short→Mealy→Pasty→Gummy
Viscosity		Thin→Viscous
Elasticity		Plastic→Elastic
Adhesiveness		Sticky→Tacky→Gooey

Geometrical characteristics

Particle size and shape	Gritty, Grainy, Coarse, etc.
Particle shape and orientation	Fibrous, Cellular, Crystalline, etc.

Other characteristics

Primary parameters	Secondary paramters	Popular terms
Moisture content		Dry→Moist→West→Watery
Fat content	Oiliness	Oily
	Greasiness	Greasy

Reprinted from *Journal Food Science* (1963), **28**, 385–9; copyright © by Institute of Food Technologists.

the model is linear in all the perceived attributes, the optimal in each relating to preference was not possible.

Later, Horsfield & Taylor (1976) showed how perception and acceptability could be linked. Using attributes agreed by group discussion, a panel was trained to score the texture and flavour of meat products. Eleven attributes were used, as shown in Table 2.

Thirteen examples of meat and meat-like products were put before the panel. The results can be considered as 13 points, in 11-dimensional space – which is clearly unmanageable! The technique of principal components analysis was used and showed that internal correlation of parameters

Table 2. *Factor loadings for each dimension (from Horsfield & Taylor, 1976)*

Sensory parameters	Dimension 1	Dimension 2	Dimension 3
Resistance	0.106	0.302	0.938
Resilience	0.242	−0.022	0.964
Initial juiciness	0.943	−0.284	0.076
Meat flavour	0.273	0.922	0.222
Soya flavour	−0.192	−0.929	0.026
Other off-flavour	0.109	−0.868	−0.253
Breakdown	0.746	0.347	0.555
Uniformity	−0.712	−0.359	−0.579
Chewiness	0.459	0.148	0.856
Final juiciness	0.931	0.223	0.254
Bolus formation	0.916	0.269	0.267

From Horsfield & Taylor (1976), with permission from the Society of Chemical Industry.

allowed the data to be 'mapped' into three complex dimensions without loss of significant discrimination between the 13 samples (Fig. 1).

The dimensions have been named by examination of the attributes having the highest correlation with each axis. These factor loadings are shown in Table 2 – from which it appears that the first dimension (on which the majority of the sample are discriminated) relates to juiciness and the formation of a swallowable bolus. Hence the term succulence was chosen for this dimension. Likewise dimension 2 was clearly related to 'flavour' and dimension 3 to 'toughness/tenderness'. To obtain 'acceptability', untrained consumers were asked to score the same samples. Contours could be constructed on the texture map, the projections of which are shown in Fig. 2. The optimum corresponds to a tender, juicy, highly flavoured steak. Again, the model was approximately linear in each of the dimensions. Presumably there are textures in meat which are too soft, juicy and highly flavoured to be acceptable as carcase meat, but these were not presented in this experiment.

Stimulus and textural response

So far, we have considered the results of textural perception by people and its reconciliation with their judgment of acceptability or preference.

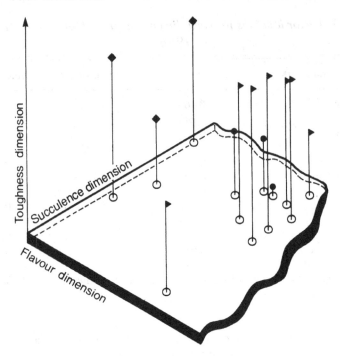

◆ Textured vegetable protein foods
▶ Carcase meats
● Processed meats

Fig. 1. Three-dimensional plot of the ten meat products and three textured vegetable foods. (From Horsfield & Taylor (1976), with permission from the Society of Chemical Industry.)

Now we need to consider the physical stimuli by which perception is triggered. Unfortunately, since the stimuli arise during the mastication process in a closed mouth, direct observation is very difficult, and previous chapters discuss methods of direct and indirect measurements of the process with different foods. One conclusion is obvious. Since mastication is a dynamic process in which the properties of the material are changing, texture perception must be itself a dynamic process relating to all of the structures present during chewing. This was formally stated by Drake (1974) (Fig. 3).

The structure of a food relates to that architecture which can be sensed mechanically during chewing. This, whilst determined by the molecular organisation, probably relates to particles and properties of colloidal and

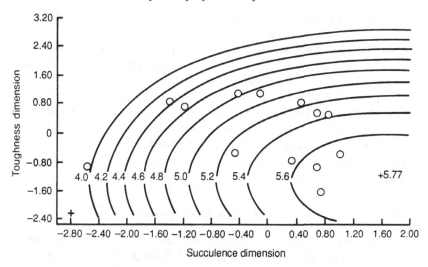

Fig. 2. Consumer acceptability contour map: O, original samples; +, the optimal product. (From Horsfield & Taylor (1976), with permission from the Society of Chemical Industry.)

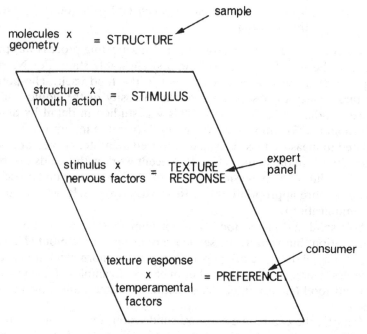

Fig. 3. Texture perception. (From Drake, 1974.)

microscopic dimensions. The product of mouth action on the structure produces the physical stimuli.

These stimuli are sensed by the tongue, lips and soft palate and interpreted after neural transmission into texture response. It is not surprising, therefore, that physical tests on the intact food structure relate only to the initially perceived textural attributes, such as resistance, initial hardness and initial moistness or juiciness. Even the early work by Szczesniak *et al.* (1963) requires more than one mechanical deformation of the food material to obtain a correlation between physical and sensory properties.

The limitations of instrumental methods in defining texture have been reviewed (Sherman, 1975).

A dynamic model of textural perception

A general model for textural perception was outlined recently (Hutchings & Lillford, 1988). We considered that such a model should include the following:

1. The behaviour of food under all eating conditions.
2. The description of the breakdown pathway during mastication.
3. Variations in the pathway caused by the food, the eater and the occasion.

We considered that the target for the masticating process was production, in the mouth, of materials in a swallowable state. For Newtonian liquid foods, the property is inherent in the food itself. The perceived texture relates to observations of the viscosity of the liquid under the shear conditions in the mouth. This was studied in detail by Shama & Sherman (1973), who showed that viscosity in the mouth was not simply related to measured viscosity and produced a 'master curve' for in-mouth viscosity of simple liquids (Fig. 4). Recent work with liquids exhibiting a yield modulus has suggested that the dynamic viscosity under oscillatory shear is more appropriate (S. B. Ross-Murphy & E. R. Morris, personal communication).

Soft solid and brittle foods are not immediately swallowable without mastication, but what is the sequence of events in the mouth? To obtain experimental data, we asked panellists to chew normally but expectorate samples after a controlled number of chews. Examples of a high moisture content food (steak) and a dry structure (crackers) are shown in Figs. 5 and 6.

Clearly, the common events are subdivision, moistening and reassembly. Panellists reported that they were at the point of swallowing at, or

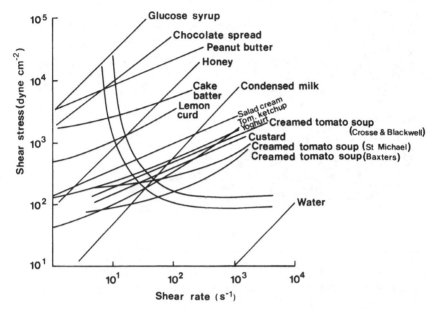

Fig. 4. Shear stress–shear rate data for food samples superimposed on 'master' curve. (From Shama & Sherman, 1973.)

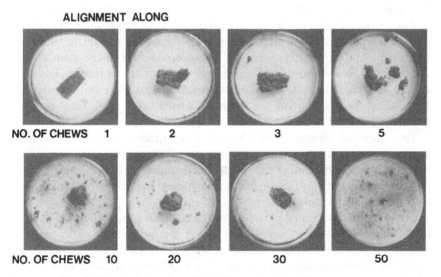

Fig. 5. Chewed samples of latissimus dorsi muscle.

COARSE CHEWER

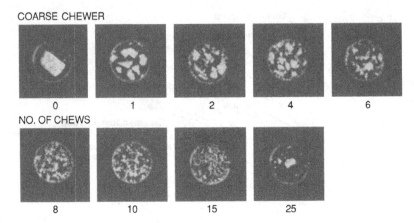

Fig. 6. Samples of Krackawheat cracker biscuits.

immediately prior to, the reassembly stage. As a first attempt at a general model of food breakdown, we propose a three-dimensional representation, where the axes are as follows:

1. Degree of structure (subdivision).
2. Lubrication (moistening).
3. Time in the mouth.

In principle, the first axis is self-explanatory and measurable by size and shape distribution analysis. The second axis is more subtle and relates to release of liquid from the food; absorption of saliva, melting or solidification of oils and fats; and the barrier properties of the latter in limiting moisture transfer. Finally, time can be approximated to the number of chews for a solid food of approximately ambient temperature. For ice cream and chocolate, rates of heat transfer must also be considered because of their effect on the 'degree of structure' axis. This 'mouth processing' model is shown schematically in Fig. 7.

The previous examples of chewed food qualitatively support the model, but can the dimensions be quantified? We have some evidence that this is possible, though the results are tedious to obtain. In Fig. 8 we show the results of wet-sieving boluses produced by increasing number of chews by the same subject.

The size distribution of particles produced from food structures of the same type (fibrous meat and an analogue) are measurably different, as were their sensory properties. As yet we have been unable to obtain a quantitative measurement of the succulence or lubrication axis with time. This dimension relates not only to the collapsing structure of the food but

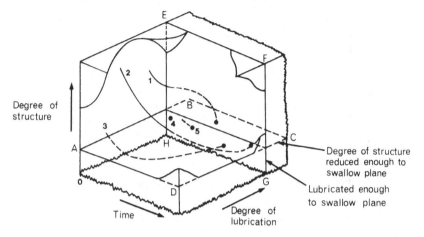

Fig. 7. The mouth process model. 1, Tender juicy steak; 2, tough dry meat; 3, dry sponge cake; 4, oyster; 5, liquids. (From Hutchings & Lillford, 1988.)

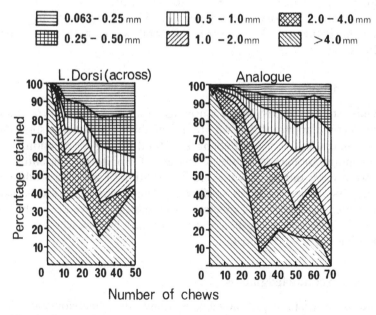

Fig. 8. Cumulative breakdown patterns for beef latissimus dorsi fibres lengthways, and analogue.

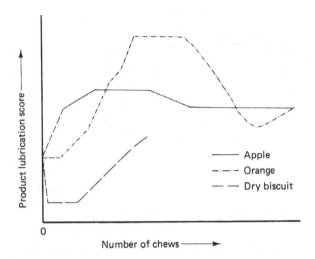

Fig. 9. Change in perceived product lubrication during progressive chewing. The point at zero chews indicates initial mouth lubrication. (From Hutchings & Lillford, 1988.)

also its demand on, and response to, saliva production. None the less, time-resolved sensory measurement of different foods clearly indicates that the pathways taken towards the swallowable state are significantly different (Fig. 9).

Conclusions

Whilst none of us pays much conscious attention to the texture of food, it has become apparent that we are all capable of measuring sophisticated mechanical and structural changes in food during mastication. Texture perception appears to be the procedure of monitoring the entire breakdown process during chewing and the addition of saliva to produce lubricated assemblies ready for swallowing. When this perceived breakdown path does not match the expectation of the food, derived from its description or visible structure, then acceptability will be low, whether or not the material is edible or nutritious.

Acknowledgments

The author gratefully acknowledges stimulating discussion and collaboration with Unilever colleagues both past and present.

References

Abelson, R.P. (1981). Psychological status of the script concept. *American Psychologist*, **36**, 715–29.

Ball, C.O., Clauss, H.E. & Stier, E.F. (1957). Factors affecting quality of prepackaged meat. Physical and organoleptic tests. A. General introduction. B. Loss of weight and study of texture. *Food Technology*, **11**, 277–83.

Davis, J.G. (1937). The rheology of cheese, butter and other milk products. *Journal of Dairy Research*, **8**, 245–64.

Drake, B. (1974). A comprehensive formula for the acceptance of food texture and its generalization to overall food acceptance. *Journal of Texture Studies*, **5**, 109–13.

Harries, J.M., Rhodes, D.N. & Chrystall, B.B. (1972). Meat texture. 1. Subjective assessment of the texture of cooked beef. *Journal of Texture Studies*, **3**, 101–14.

Horsfield, S. & Taylor, L.J. (1976). Exploring the relationship between sensory data and acceptability of meat. *Journal of the Science of Food and Agriculture*, **27**, 1044–56.

Hutchings, J.B. & Lillford, P.J. (1988). The perception of food texture – the philosophy of the breakdown path. *Journal of Texture Studies*, **19**, 103–15.

Schutz, H.G. & Damrell, J.D. (1974). Prediction of hedonic ratings of rice by sensory analysis. *Journal of Food Science*, **39**, 203–6.

Shama, F. & Sherman, P. (1973). Identification of stimuli controlling the sensory evaluation of viscosity. II. Oral methods. *Journal of Texture Studies*, **4**, 111–18.

Sherman, P. (1975). Textural properties and food acceptability. *Proceedings of the Royal Society of London*, Series B, **191**, 131–44.

Smith, H.R. (1947). Objective measurements of quality in foods. *Food Technology*, **1**, 345–9.

Szczesniak, A.S. (1963). Classification of textural characteristics. *Journal of Food Science*, **28**, 385–9.

Szczesniak, A.S., Brandt, M.A. & Friedman, H.H. (1963). Development of rating scales for mechanical parameters and correlation between the objective and sensory methods of texture evaluation. *Journal of Food Science*, **28**, 397–403.

Index